INFORMATION OPE
FACTS FAKES CONS

All royalties from the sales of this book will go to Hounds for Heroes. Their mission is to provide specially trained assistance dogs to injured and disabled men and women of both the UK Armed Forces and Emergency Services.

Information Operations
Facts Fakes Conspiracists

STEVE TATHAM

Copyright © 2024 Steve Tatham

First published in 2024 by
Howgate Publishing Limited
Station House
50 North Street
Havant
Hampshire
PO9 1QU
Email: info@howgatepublishing.com
Web: www.howgatepublishing.com

All rights reserved.

No part of this publication may be reproduced, stored in a retrieval system, or transmitted in any form or by any means including photocopying, electronic, mechanical, recording or otherwise, without the prior permission of the rights holders, application for which must be made to the publisher.

British Library Cataloguing-in-Publication Data
A catalogue record for this book is available from the British Library

ISBN 978-1-912440-47-4 (pbk)
ISBN 978-1-912440-46-7 (hbk)
ISBN 978-1-912440-60-3 (ebk - EPUB)

Steve Tatham has asserted his right under the Copyright, Designs and Patents Act, 1988, to be identified as the author of this work.

The views expressed in this book are those of the author and do not necessarily reflect official policy or position.

Contents

Foreword   *vii*
Preface   *xi*

Introduction   1
1   What Are Information Operations?   11
2   Britain and IO: From Eunuchs to the Archers   32
3   The USA – Bigger and Better?   51
4   Iraq and Afghanistan   63
5   Understanding Audiences   88
6   Past IO Campaigns   96
7   Russia and Ukraine   117
8   Iran, North Korea and China   139
9   It's All About Psychology   161
10   Conspiracy Theories and Useful Idiots   175
11   The Cambridge Analytica 'Conspiracy'   190
12   The Future?   214
13   Conclusions   232

Index   *249*
Previous Publications   *255*

Foreword

The US and its NATO allies have the world's best Armed Forces. The best trained soldiers, sailors and aviators wear British, US, and uniforms of other NATO nations. They carry, train on and fight with the most advanced equipment and weapons. But combat prowess on physical battlefields is no longer sufficient to deter conflict and, if necessary, to win wars. That is because adversaries and potential enemies are developing other means to achieve their objectives. Those means include Information Operations (IO) aimed at exerting influence over and controlling the information environment. Authoritarian powers Russia and China have become effective at employing a range of means to control their own information environments while using cyber-enabled information warfare including disinformation and propaganda to undermine the cohesion and will of our free and open societies. Our adversaries know that IO don't just support conventional military operations, they have become a conventional means to accomplish strategic objectives below the threshold of what might elicit a concerted military response.

Vladimir Putin has used Information Operations as a way to accomplish his objective of restoring Russia to national greatness on the cheap. Putin understood that Russia, with a GDP smaller than Italy's, could not compete with Europe and the United States directly. His theory of victory is to help other nations drag themselves down so that he might be the 'last man standing'. Russian IO are part of sophisticated campaigns designed to sow divisions in society, cloud reality, stoke conspiracy theories, and, above all, create crises of confidence in democratic principles, institutions, and processes.

China has learned from Russia and in some ways has surpassed the Kremlin's ability to create crises of confidence with overt 'Wolf Warrior Diplomacy', and more subtle means of influencing perception and behaviour abroad through social media algorithms and disinformation that portray China as a victim rather than a perpetrator of aggression and spans cyberattacks, economic coercion, and physical attacks from the Himalayan frontier with India to the South China Sea to Taiwan to Northeast Asia. The Chinese Communist Party (CCP) leadership is motivated, in part, by fear of losing their exclusive grip on power. That is why the CCP has prioritized information warfare against its own citizens.

Information Operations are foundational to the CCP's efforts to perfect a technology-enabled Orwellian police state. China's 'great firewall' keeps out any information or content that might challenge its narrative, expose its egregious treatment of its own population such as the slow genocide of the Uighurs, or encourage the Chinese people to want a say in how they are governed. Ubiquitous surveillance and monitoring combined with artificial intelligence technologies allow the Party to pre-empt opposition. The CCP's social credit score system weaponizes people's social networks against them and restricts access to basic services for those who engage in even mild criticism of the party-state. Meanwhile all forms of media to which people have access extol the virtues of the Party and credit its leaders with all that is good while pushing narratives of Western decadence and decline.

Terrorist, insurgent, and criminal organizations also understand the importance of IO. These groups often combine violence with sophisticated propaganda to gain the support or acquiescence of the population. It is for this reason that the counterinsurgent's and counterterrorist's success against these groups requires efforts to clarify their own intentions while countering enemy disinformation. Ideas are often weaponized in the information age and that is why Information Operations are and will remain integral to all conflicts.

As the author makes clear, we have work to do. As I write this foreword, citizens of many nations across the global south have been subjected to sustained campaigns designed to advance the interests of authoritarian regimes; disparage the United Kingdom, the United States and other democracies; and undermine confidence in democratic principles, institutions, and processes. Successful IO is one of the reasons why many nations have failed to condemn or have provided tacit support for Putin's criminal war in Ukraine. Putin's IO in the Middle East, in Southeast Asia and in Africa have gone largely unchallenged. In Russia, despite tens of thousands of dead Russian soldiers being sent home, Putin's internal IO have largely stifled dissent. In China, no one today discusses Tiananmen Square or deigns to question the wisdom of CCP leaders. As in George Orwell's book, 1984, inconvenient history is erased. The events of April 1989 and the deaths of so many protestors just 'never happened' - neither did the Great Leap Forward or the Cultural Revolution during which the CCP killed more people than dictators Joseph Stalin and Adolph Hitler combined. In the Chinese information space, the CCP is the cause of everything good and foreign forces are the source of all problems. Information Operations distort reality to the extent of collective brainwashing in totalitarian North Korea. Other authoritarian states, from Venezuela to Cuba to Zimbabwe to Cambodia have invested heavily in the ability to control and distort information. These regimes share techniques and continuously improve their ability to shape and police the thoughts and behaviours of their people.

How do free and open societies compete on the battlegrounds of information and perception? In the months after the mass murder attacks of September 11, 2001, some leaders looked to the advertising and marketing worlds to improve communications between the Western democracies and the Muslim world. The 'ad exec' was favoured over psychologists and social scientists; attitudes and perceptions afforded primary over behaviours; opinion polling over proper behavioural research. With few exceptions, the effort to craft and deliver effective 'messages' to target populations failed as those crafting those messages were unfamiliar with the political, social, religious, and psychological dynamics influencing behaviour in places like Afghanistan and Iraq.

The latest trend in the global IO community is to focus on social media. Now the 'IT exec', is granted priority over psychologists and social scientists. And the narrow focus on attitudes and perceptions remains disconnected from the behaviours that drive conflict and insecurity. That is why this book is appearing at the right time and why it deserves a wide readership.

Steve Tatham is the right person to improve our competence in Information Operations. I first met him in Kabul, Afghanistan as he was trying to reorient Psychological Operations on destructive behaviours instead of perceptions and attitudes. He helped the multinational command understand better what motivated people to act in ways that were perpetuating violence and insecurity. In this book Steve quotes a line from the Hollywood film, Miracle on the Hudson. "Can we get serious now? We've all heard about the computer simulations, but I don't quite believe we still have not taken into account the human factor". It's a great quotation that summarizes Steve's main point: the Human Factor must have primacy in IO. If we don't understand the human factor IO will fail to produce the desired results. People's online personas and social media scribblings are less important than human behaviour.

Tatham has given us the opportunity to reappraise IO. We learn what it is, who to target, how to understand different audiences, what to say or do, and how to measure success. Whether we can build a better future in the coming century and how much blood will be shed will depend, in part, on how well the UK, the US, and our allies operate in the information space. Reading, discussing, and implementing the ideas in this book is a great starting point.

Lieutenant General H.R. McMaster US Army (rtd)
Former US National Security Advisor
Stanford, California USA

Preface

When this book is published, I will have been an Officer in the Royal Navy (RN), Active Duty, and Reserve, for over 35 years. Of that, nearly 25 years has been spent in the world of Information and Psychological Operations. In the UK, at least, there are probably no more than a handful of officers who have similar longevity and experience and hardly any of them have risen above the rank of Colonel or their Royal Air Force (RAF) and RN equivalents.

In 2004 I wrote my first book, *Losing Arab Hearts and Minds*, which told the story of the failed battle for Arab opinion in the 2003 invasion of Iraq. As one of the UK's public spokespeople, and with special responsibility for Arab Media, I felt I was well qualified to comment and the book's title reflected my view of our structural inability to understand population groups and what to say or do to influence them. Post my tours in Afghanistan I and General Andrew Mackay, in my opinion one of the most successful Generals this country has ever produced, wrote Behavioural Conflict. General Stanley McChrystal, the famed US Commander, agreed to write its foreword. Like my first book, Behavioural Conflict was written on the back of operational experience, this time in Afghanistan. It is my personal view (and certainly not that of the UK MoD) that one of the most significant reasons for the failure and ignominious pull out from Kabul in 2021, was that we failed to understand Afghans and where we spent huge amounts of effort and money trying to change Afghan attitudes and perceptions we should have been focusing on understanding the drivers for their behaviours and the influence that the Taliban had over them.

This book, which is based on the totality of my experiences, is my last. In many respects, being an Information Operations officer is a frustrating experience. Because, in the UK at least, it is not seen as a military profession, everyone has a view on what good Information Operations looks like. Through the years countless senior officers have passed briefly through the world of Information Operations. Mackay, and a small number of others, were an exception to the vast bulk who arrived believing they intrinsically understood the nature of Information Operations and proved impervious - sometimes even resistant - to advice or experience. For a long time I was a regular guest lecturer at the UK's Joint Information Activities Group – the UK's 'schoolhouse' for Information Operations – and a recurring conversation with the staff there was how difficult it was to educate on the importance of Information Operations. I first felt that in 2003; that the same issues remain over 20 years later is disheartening, especially given our collective problems in Iraq and Afghanistan and the now daily reports from Ukraine and Gaza where information is being deployed on both sides for operational advantage.

In 2018 I became unexpectedly embroiled in the Cambridge Analytica affair. It was one of the most stressful events of my life. Suddenly I was the focus of global social and mainstream media attention; often devoid of much factual basis, allegations were made that my work in Information and Psychological Operations had led to the election of President Trump and Britain's exit from the European Union (BREXIT). Daily I would read more and more Tweets, social media posts and newspaper articles that linked my work in defence with the US election and with the clear dishonesty that revolved around the Leave campaign in the UK European referendum. If I were a US citizen, I would probably vote Democrat but certainly never Trump. As a Brit I am a European who voted 'remain' and views BREXIT as a tragedy and a national disaster. That I should somehow be accused of behaviours that were the anthesis of my beliefs and values was deeply upsetting.

In the face of daily attacks, I sank into deep depression. I was privileged, however, that as a Reserve Officer I had incredible employers, The Royal Navy. The Navy asked me to explain the facts behind the headlines; once they understood what had really happened, and not what people imagined happened, they stood by me, unequivocally, and one hundred per cent. Three senior officers in particular, Commodore's Gary Sutton, Martin Quinn and Captain Paul Hill, told me that they had my back. And by goodness they did. Their collective advice was that I should not enter into the public debate, which meant not rebutting or disputing any of the stories, however ridiculous, as to do so would prolong the affair. It was good advice, which I followed, but it was painful to read often inaccurate and unpleasant comments about oneself each day. However, what I was experiencing was nothing to what others, good people, were going through. I do not know how Nigel Oakes, a dear friend, and a visionary in the work of Information Operations, survived. Indeed, he nearly did not, the stress and constant attacks accelerating his cancer and, at one point, seeing his family called to his hospital bedside to say goodbye. Thankfully, miraculously, he pulled through but for ever physically and emotionally weakened. Others, good people who had worked hard to support defence and had nothing to do with Cambridge Analytica, kept looking over their shoulders and wondered if and when 'the mob' would come for them.

I got through it because of the support of my wife, family, and incredible friends – not least General Andrew Mackay who was an absolute rock. But I also got through because on the worst days my two dogs were always there for me, offering unconditional love and support. Which is why Hounds for Heroes is my chosen charity and to whom all author's profits will be donated. Dogs never judge. Dogs somehow know when you are down and always manage to find a way to comfort or lift spirits. Allen Parton, the founder of Hounds for Heroes, knows this too. And he has made it his life mission to bring dogs to the lives of veterans, military and blue light, who are struggling. I have had the opportunity to talk to some of them and can see the strength they take from their four-legged friends. Please read Allen's story and, if you can, support them in their amazing work.

This book, the culmination of my Information Operations career, is intended to be read by anyone. I have made it neither too academic nor too 'military'. I hope the various anecdotes and stories make it an enjoyable and accessible read. In putting it together I am indebted to some wonderful people who freely gave so much of their time and expertise.

Mr Matt Armstrong, the former Executive Director of the U.S. Advisory Commission on Public Diplomacy, was a good enough friend to tell me where I had got it wrong and how to improve. His advice, particularly in my discussion of the US, was invaluable and I am indebted. So too Colonel Ian Tunnicliffe who once ran the MoD's Department for Information Operations, Colonel Tony Paz the former Commanding Officer of U.S. PSYOP, Major Collins Devon Cockrill of the U.S. Army Reserve PSYOP, long standing Russian disinformation specialist Mr Todd Leventhal of the US State Department and Gaby Van Den Berg who helped me with the Dutch references, 'R', a former UK Special Forces Officer and invaluable devil's advocate and General Richard Barrons, who supported me both during his illustrious military career and subsequently in his business one. My great friend, and former co-author, General Andrew Mackay nearly three years ago waded through a discarded manuscript and told me I needed to get on and finish it. Fellow Naval Officer Frank Ledgwidge wrote the wonderful 'Losing Small Wars' and Dr Mike Martin, a fellow veteran of Afghanistan and strategist extraordinaire both offered invaluable support.

So too the UK's foremost Russia watcher, Keir Giles. My friend, wounded veteran Colonel Dr Chris Naler of the US Marine Corps, provided steadfast support – to him I say, 'Latvernergo' (which will mean nothing to anyone else but will bring a smile to his face in the Texas sun).

To General HR McMaster, an incredible thinker and inspirational leader whom I first met in Kabul, thank you for your care, time and interest in writing the Foreword. To Kirstin Howgate, my publisher, who has taken me, and this manuscript, on, thank you. The world of publishing is mercenary, hard and unfriendly; working with you has been an absolute pleasure. There are a number of other people to whom I am indebted but who prefer to remain anonymous. I am also hugely grateful to the many friends on both sides of the Atlantic who have given their time and help freely and enthusiastically. I am humbled by such friendship.

Finally, to my sanity checker and long-term proofreader, Major Ric Cole Royal Marines. Like me, Ric is a long time Information Operations practitioner and comrade in arms. He, like me, has experienced the highs and lows of Information Operations and he has also been a keen tutor and mentor to younger officers. It is they who are one of my key target audiences – young men and women who today, astonishingly, will be flung into vital Information Operations assignments with little support and who will do their best in the face of often inexperienced Information Operations leadership and unrealistic expectations. I hope this book helps them.

Because so much of Information Operations is visual or audible a website has been constructed to support this book. That website, **www.IOFFC.info**, contains pictures, videos, sounds and imagery that supports each book chapter and I hope, will in time, become a helpful reference site for Information Operations.

Disclaimer: Although I am still a member of the Royal Naval Reserve, the opinions expressed in this book are a reflection of my indepedent academic and consultancy career and follow on from my past books and publications in Information Operations. They are not intended, in any way, to reflect the views of the UK Ministry of Defence or the Royal Navy and this publication is not endorsed by either.

Steve Tatham
New Forest, Hampshire, UK
2024

Introduction

It was difficult to determine their ages, but the children were young, some no more than seven or eight, the oldest no more than 15 or 16 years old. Chasing after the football, screaming at the top of their voices with excitement and happiness, they could have been kids anywhere. But when you got a little closer you realised they weren't kids from anywhere; some had deep scars on their arms, forming the letters R, U and F. Others were amputees; some with hands cut at the wrists, others at the elbow, macabrely known locally as 'long' or 'short' sleeves. All of them were orphans. Victims of a vicious civil war in the former West African British colony of Sierra Leone.

British sailors joined in the game, but some hung back to take photographs and talk to the few brave adult staff who did their best to look after the children. They heard their pitiful stories, they heard about the rapes, murders, and cruel violence inflicted indiscriminately upon the kids' families, on men, women, and children. And from the shadows of the bush, from the darkness of the jungle which surrounded the makeshift pitch, others watched; too frightened to come into the light; suspicious of everyone and everything. Later that day the sailors returned to their ship, the Royal Navy aircraft carrier HMS Illustrious, and posted their pictures in the passageways and mess decks. This was why the ship and her 1200 crew were here; this was why British troops were in the country; this was why we had joined the British Armed Forces.

It's May 2000. The Fleet Flagship, HMS Illustrious, sits a few miles off the coast of Sierra Leone. Inland the Army of the Revolutionary United Front (RUF) has advanced on the ramshackle city of Freetown, capital of the small West African state. Since 1991 Sierra Leone had been engulfed in almost constant civil war – a victim of corruption, poverty, poor political leadership and, perhaps more than anything else, the country's abundant supply of diamonds buried deep in the nation's heart beneath the red verdant African soil. In an attempt to bring peace to the country the year before the Lome Peace Accord had been signed[1] which offered a power sharing arrangement between the existing Sierra Leone Government and the RUF and, to help police it, established the United Nation's Mission to Sierra Leone (UNAMSIL).[2]

Despite the presence of UN Peacekeepers, the country had lurched from violent crisis to violent crisis. The RUF, led by former Sierra Leone Army Corporal Foday Sankoh, and supported and funded by Charles Taylor, leader of the neighbouring state of Liberia, had embarked upon a vicious civil war the currency of which was mass rape, indiscriminate killing and amputations. Between 1991 and 1999 the war has claimed over 75,000 lives, caused half a million people to become refugees, and displaced over half of the country's 4.5 million people.[3] Built largely upon a child army, the RUF abducted and drugged boys as young as seven years old, supplied them with weapons, trained them to kill and began a reign of terror across the rural heartlands of the country. Targeting civilians, especially women and children, they used amputation as their choice of terror: feet, hands, ears, lips, and noses were all removed by machetes and, by May 2000, the capital was heaving with refugees, the ill, injured and displaced, many of them camped out around the British High Commission and U.S. Embassy.

Responding to a request from the UN Secretary General, the UK Ministry of Defence deployed its Joint Task Force Headquarters (JTFHQ) to Freetown on 5th May 2000 to prepare for an evacuation operation of British, Commonwealth and European citizens. An Operational Liaison and Reconnaissance Team (OLRT) was deployed the same day to Freetown and on 7th May the Royal Navy's Amphibious Ready Group (ARG) was ordered to sail to Sierra Leone to assist, arriving six days later. Concurrently the Royal Navy aircraft Carrier HMS Illustrious was diverted from Exercise Linked Seas in the eastern Atlantic and arrived off the coast of Sierra Leone on 11th May with me embarked.

Barely more than one year previously the UK's Prime Minister, Tony Blair, had delivered his so called 'righteous intervention' speech in which he had outlined a doctrine where the international community

---

1 Peace Agreement between the Government of Sierra Leone and the RUF (Lomé Peace Agreement), United Nations Peacemaker, available at https://peacemaker.un.org/sierraleone-lome-agreement99.
2 United Nations Mission in Sierra Leone. See https://peacekeeping.un.org/mission/past/unamsil/.
3 Smillie I, Gberie L, Hazleton R. *The heart of the matter – Sierra Leone, diamonds and human security*. Partnership Africa Canada; 2000, available at https://reliefweb.int/report/sierra-leone/heart-matter-sierra-leone-diamonds-human-security.

could justify intervention, including if necessary military intervention, not just when a nation's interests were directly engaged but also where there existed a humanitarian crisis or evidence of gross oppression of a civilian population.[4] Both these latter criteria were present in Sierra Leone and both were undoubtedly strategic drivers for the UK presence. However, as the mission evolved the international political landscape changed, particularly when other West African nations – notably Liberia – accused the UK of neo-colonial imperialism. In the UK the mission caused some concern and accusations of 'mission creep'[5] began to emerge in UK papers. With the strategically important airfield at Lunghi secured by British troops and the UN headquarters at the Mammy Yoko hotel on a small peninsular off Freetown also safe, the UK began reducing its military profile. Rules of Engagement were changed and anything that might feed the narrative that the UK was an aggressive actor was curtailed. HMS Illustrious had over 20 Harrier aircraft embarked yet even basic overflight flights were heavily curtailed.

With the evacuation complete there was significant pressure for British troops to leave but the commander on the ground, Brigadier David Richards, was concerned that the UN Forces would be vulnerable to attack and that the RUF might yet overwhelm Freetown. The consequence for the population would be catastrophic and a massacre, not unlike in Rwanda, was possible. And just like in Rwanda, the United Nations could again be humiliated. British Forces therefore continued to conduct protective operations, securing key locations, junctions and roads and Harrier jets began low-flying operations (carrying weapons for self-defence only) to try and dissuade an RUF advance. On 12th May 2000 the UK Government announced that British Forces would remain in Sierra Leone to support UNAMSIL and in June Operation BASILICA was established – a training and mentoring mission for the Government of Sierra Leone (GoSL) Armed Forces. On 10th Nov 2000 a permanent cease-fire was negotiated.[6] However, like so many before the RUF largely ignored it and began pushing again

---

4 Speech by British Prime Minister Tony Blair to the Chicago Economic Club, 22 April 1999, available at https://www.chicagotribune.com/news/ct-xpm-1999-04-23-9904230097-story.html.
5 Marcus J, *Mission Creep in Sierra Leone?*, The BBC, 19 May 2000, available at http://news.bbc.co.uk/1/hi/world/africa/755267.stm.
6 Sierra Leone: Treatment of Revolutionary United Front (RUF) prisoners upon the signing of the Lomé Accord and the subsequent ceasefire agreement; details of the amnesty granted to members of the RUF, UNHCR, available at https://www.refworld.org/docid/403dd21a8.html.

into the outskirts of the city. To demonstrate the UK's determination, it was decided to run a series of Information Operations (IO), comprising displays of military capability and power to the RUF, what one commentator described as: "a controlled and understated display which, undoubtedly proved Britain's ability to deploy troops rapidly".[7] Between 13th November and 8th December Operation SILKMAN,[8] as it came to be called, was conducted. It involved over 600 Royal Marines along with artillery and aircraft, as well as elements of the Sierra Leone Army. The exercise consisted of a series of amphibious landings, a parachute drop[9] and fire power demonstrations using multiple assets to create an image of power, posture and intent. But on cessation of the operation most of the participants returned home to the UK.

Operation PALLISER, and the follow-on operations of BASILICA and SILKMAN was, the largest purely national deployment of UK military force since the Falkland's war. It was mounted at extremely short notice, was a complete success and had an immediate strategic impact. A key component of its success was the enforced use of Information Operations; enforced because the prevailing political climate changed so quickly that conventional military firepower (so called 'kinetics') were prohibited other than in self-defence. The IO plan, which from the outset was secondary to the main task, quickly assumed primacy over the entire mission. In reviews subsequently undertaken of the operation, the UK Ministry of Defence concluded that Information Operations, whilst still immature in terms of deployment and management, were extremely effective and had resulted in the RUF's decision making cycle being overloaded with information. In other words, IO had been vital to the success of the operation.

Today the UK's intervention in Sierra Leone is often quoted by military historians as the exemplar of an overused and often trite government phrase, 'A Force For Good'. It is also widely recognised in military doctrine as an example of ingenuity and adaptability. Information Operations was not a planned component. Indeed, the operation happened so quickly that huge

---

7 Evoe, Patrick, J. (2008) *Operation Palliser: The British Military Intervention Into Sierra Leone, A Case of a Successful Use of Western Military Interdiction in a Sub-Sahara African Civil War*. MA Thesis: Texas State University-San Marcos, available at https://digital.library.txstate.edu/bitstream/handle/10877/2602/fulltext.pdf.
8 Operation Silkman: What was the last solely British military campaign?, available at https://www.forces.net/news/operation-silkman-what-was-last-solely-british-military-campaign.
9 *RAF Regiment remembers 20th anniversary of Sierra Leone parachute drop*, available at https://www.raf.mod.uk/news/articles/raf-regiment-remembers-20th-anniversary-of-sierra-leone-parachute-drop.

swathes of it had to be made up as the campaign developed. The IO staff was small and empowered by the Command. Because of this, comparatively junior officers were able to develop plans quickly, enacting them with the approval of the local commander but without in-depth scrutiny by Whitehall. This led to an agility and responsiveness that can be uncommon in other operations. So much was achieved through Information Operations by a small and largely unqualified cadre of personnel that the scope of what could have been achieved with properly resourced and trained people is potentially huge. For this reason, PALLISER and SILKMAN (which was far better developed than PALLISER) rightly hold a premier position in any assessment of contemporary military Information Operations.

Three years after the Sierra Leone conflict I was appointed head of in-theatre maritime media operations and, eventually, public spokesman for the Royal Navy in the 2003 Iraq War. This was the first time that I had ever come across the Arab media. The Al-Jazeera TV station was already a few years old, but Al-Arabiya TV had just started and the two channels had significant market share across the Arabic speaking world. Whilst I knew a great deal about the British and US media, and a great deal about sentiment in the US and UK around the coming conflict, I knew absolutely nothing about how the Arabic speaking world viewed it. It was a baptism of fire and whilst I found the journalists working for Al-Jazeera and Al-Arabia thoroughly professional and trustworthy in all that they did, I was poorly prepared for the visceral anger, and in many instances hatred, of their viewers. Whilst the Ministry of Defence directed that I and my fellow spokesmen should talk about Freedom, Liberation and Democracy these were not 'lines' that carried much weight in the Arab world. They looked hollow and Arabs not unreasonably asked if the civilian casualties – which we so coolly referred to as collateral damage – were reaping the benefits of their liberation and democracy? Just like in Sierra Leone, the UK knew almost nothing about the various population groups caught up in the conflict eco-system. I wrote about my experiences for my Master's thesis at the University of Cambridge in 2004. I had been selected as one of two Royal Navy students to attend the MPhil in International Relations. That thesis became my first book, Losing Arab Hearts and Minds, published in 2006.[10]

---

10 Tatham S, *Losing Arab Hearts and Minds: The Coalition, Al Jazeera and Muslim Public Opinion* (Hardback), April 2004, available at https://www.waterstones.com/book/losing-arab-hearts-and-minds/steve-tatham/9781850658115.

For me Sierra Leone and Iraq marked watershed career moments. Up until then military service had been very much that which appeared in the recruiting literature. Smart white uniforms, exotic locations, drinks on the ship's quarter deck, parties, and the prestige of being an officer in the world's oldest and most professional Navy. My career had spanned peaceful deployments to the far east, service in the last bastion of the British Empire, Hong Kong, and a glorious exchange to the Royal New Zealand Navy. The wars in Sierra Leone and Iraq changed everything. In Sierra Leone, in particular, the brutality and savagery of conflict hit you in the face. For a middle-class boy growing up in the peace and security of the UK, the sheer scale of the violence was shocking. Suddenly being a member of the armed forces was different. It wasn't just smart uniforms and glamourous port visits. Suddenly it was death, decay, brutality, inhuman savagery and an insight into the utter madness of conflict and war.

It was a watershed moment for another reason as well; Sierra Leone showed how Information Operations, if properly conducted, could achieve the most amazing results and reduce, perhaps even remove completely, the need for military force – or 'kinetics' in military speak. If we could properly understand why people behaved in specific, violent, ways – why they thought that amputating children's arms was 'reasonable' – then we might have a chance to deter that behaviour, to persuade them down another, less violent, path and to do that without using weapons and risking the horror of 'collateral damage' – the euphemism for the death and destruction caused when 'friendly' weapons find the wrong targets. It was the start of a personal journey, conducted sometimes as a participant and sometimes as an observer, that led through further conflicts – Iraq, Afghanistan, Ukraine, Gaza – and a raft of smaller conflicts, globally. It was a journey that coincided with the massive acceleration of the internet and the use of social media. It coincided with the ability of every voice to have a platform; of every view and opinion to find an audience and of images, news, videos and sounds to be forged and faked. It coincided with nation states institutionalising information in their strategy, of conspiracy theories and theorists being afforded platforms for the most ridiculous views and opinions. It coincided with a new era that French President Emmanuel Macron described as one in which: "the burden of proof has been reversed: while journalists constantly have to prove what they say – in accordance with the ethics of their profession, they must show what they say or write to be true – those spreading fake news shout out: "It is your responsibility to prove that we are wrong!"[11]

And along the way some elements of the media, a few academics and social media, often hunting as a pack, sought to 'expose' everything and everyone connected to it. Upset, angry and disorientated by US voters choosing Trump and incredulous that anyone would in their right mind vote for Brexit, some found the perfect outlet – 'military grade Information Operations'.[12] Meanwhile many serious commentators shook their heads, first in bemusement and, latterly, astonishment. In mainstream media but now also in social media, every action was seen through a lens; the political left lens; the political right lens; the Pro-Trump nationalist lens; the pro-Putin Russian lens; the conspiracy lens; the subversion of democracy lens. Facts became largely irrelevant, instead the currency of discourse had become outrage, emotion and a court of public opinion that will always find someone guilty of something.

The term Information Operations, and its often conflated corollary, Psychological Operations, had morphed from terms once known largely only by military personnel to terminology used in everyday by media, commentators and even the general public. And the wider its usage the corrupted and dirtied the terms became, a result in most instances of ignorance but in some of wilful and deliberate distortion.

This book is written to present the facts of Information Operations and why their use will and must expand. It will expose some of the fakes and fictions that have been attributed to military Information Operationsand it will address some of the conspiracists who have distorted and corrupted debate about Information Operations. The attack on Information Operations comes at a time when our global adversaries are stepping up their activities in the information environment. The annexation of Crimea in 2014 – a landmass the size of Maryland in the US – was achieved by Russian military forces with almost no shots fired. Russia directly interfered in the

---

11 New Year's Speech to the Press by Emmanuel Macron, President of the French Republic. Paris, 3 January 2018, available at https://www.elysee.fr/en/emmanuel-macron/2018/01/03/new-years-greetings-by-president-emmanuel-macron-to-the-press.

12 For example see *The New York Review of Books*, 'Beware the Big Five' by Tamsin Shaw 5 April 2018 issue, available at http://ringmar.net/wp-content/uploads/2022/02/Shaw-Beware-the-Big-Five.pdf. Or International Business Times, Russia Hacking NATO by AJ Dellinger 10 April 2017, available at https://www.ibtimes.com/russia-hacking-nato-troops-near-russian-border-had-phones-compromised-2597378.

2016 Presidential election[13] and may well have done so in the UK Brexit referendum.[14] During the 2022 invasion of Ukraine, Information Operations were used extensively by both sides. Iran has become adept at exporting not just its ideology and its Republican Guards to friendly countries – the latter allegedly in a base in Venezuela now just 2000 miles from Texas[15]– but so too their pernicious Information and Psychological Operations. China is using information to further its economic and territorial interests in the pacific region and across Africa; so too North Korea, which may have the reputation of being the hermit kingdom but is clearly not asleep when it comes to Information Operations. Al-Qaeda is not dead, nor ISIS, and both used information to good effect in their battles for a global caliphate. And in Gaza, as this book readied for press early 2024, the cause of the explosion at al-Ahil hospital on 17 October 2023 remains hotly contested.

This book is designed to challenge some of the distortions about Information Operations and explain why it will become ever more critical to success in the future conflicts that are undoubtedly coming. When undertaken properly, Information Operations can protect our democracies and their institutions; Information Operations can defend populations against the insidious effect of fake news. When done badly or not at all it risks defeat at worst and condemnation at best.

Sources. Where possible the source of information is provided in the footnotes. I believe this is important for transparency and it allows the reader to check the reference and form their own view. Many are taken from newspapers who rightly might be accused of having their own specific agendas. I have tried to substantiate each reference, where possible finding other corollary references; this has not always been possible and therefore I leave it to the reader to form their own view on the basis of the evidence presented. However, military Information Operations tend, rightly or wrongly, to be highly classified. Where public reference to such operations is possible it is included but many examples are not in the public domain and are highly unlikely to be so for many years come. This is because successful Information Operations can be replicated in other environments

---

13 *Russian Interference in 2016 US Elections*, FBI, available at https://www.fbi.gov/wanted/cyber/russian-interference-in-2016-u-s-elections.
14 Ruy D, *Did Russia Influence BREXIT*, Centre for Strategic and International Studies, 21 July 2020, available at https://www.csis.org/blogs/brexit-bits-bobs-and-blogs/did-russia-influence-brexit.
15 Carroll R, *Iran's elite force expanding influence in Venezuela, claims Pentagon*, The Guardian, 27 April 2010, available at https://www.theguardian.com/world/2010/apr/27/iran-venezuela-pentagon-report.

and the tactics and procedures that are used deserve to remain confidential in case they impact on current operations. It is also the case that some military Information Operations have been instigated for the long term, indeed variants of them may even still be underway. Finally, the details of many Information Operations simply do not exist – either because their classification at the time was so very sensitive that their knowledge was confined to a very small circle of people – known as compartmentalisation – or that they were conducted alongside allies, and in particular the US, who might veto any release of information.

This is always a frustration to historians but as someone on the inside of many UK and US military Information Operations for some years I have sympathy for such rules and I am also mindful that signing the UK Official Secrets act places in many instances a life time of obligation. Where Information Operations are explained, but without reference, they are done so having interviewed some of the officers and officials behind them who in many instances do not wish to be named for very good reasons. For example, the Salisbury poisonings illustrate the long and vindictive memories that Russian authorities keep against these who worked against them. It would be naive to presume that Information Operations have not been deployed against Russia, but it would be equally naïve, quite possibly dangerous, to detail them here. Another example is Operation KINGFISH, which resulted in extremely violent drug dealers being sentenced to long but not lifetime prison sentences; the fear of retribution cannot be ruled out. To an extent therefore I, and the many officials who have contributed, have exercised a degree of self-censorship and, since I remain a serving Officer in the Reserve Forces, submitted the manuscript to UK authorities to ensure that the Official Secrets Act was not broken.

This should not be regarded as a cover up – were that the case this book would not have been written at all – but respect for both the law and our own national security. I have never taken the view that the public has the automatic right to unrestricted access to information surrounding national security. Ta-Nehisi Coates, a former national correspondent for *The Atlantic* magazine, perhaps sums it up best:

> The very general proposition of journalism is this: The public has a right to know true things that are important to the public. It is the job of journalists to supply the public with these true things. This broad idea applies in practice not just to the goings-on of government, but

> to crime, and business, and science, and sports, and the actions of all sorts of people who are famous and/or notorious, either temporarily or permanently. But "newsworthy," a term that could be applied to everything from Watergate to sex tapes, lacks the moral force of claiming to act on behalf of the presumed rights of the public. "Newsworthy" describes how journalism works. But it doesn't engage the complicated, constant ethical dilemmas which journalists face over what to report and what not to report. I get nervous when I see journalists blithely and casually invoke the right of the public to know, without any attempt to define those terms, their limitations, and their history.[16]

Having served in uniform for thirty-six years, I am firmly of the view that democracies can as easily be impaled by freedom of information as strengthened and that the general public, academics and journalists are rarely well equipped to understand the totality of what is at stake. I am also firmly of the view, born from hard experience, that conspiracies only very rarely have any truth to them.

---

16 Coates T, *On the Right to Know Everything*, The Atlantic, October 2016, available at https://www.theatlantic.com/notes/2016/10/on-the-right-to-know/502916/.

1 What Are Information Operations?

> *"They whisper about you ... with so many leaks you probably think it could be anyone. Donald, it's everyone".*

On 7th July 2020 at seven o'clock in the morning in Washington DC, a 1 minute 17 second video was released on the Twitter (now 'X') social media platform. Within 25 minutes it had been retweeted 4,500 times. By 8am that had more than doubled and it had been viewed nearly 500,000 times. An hour later, as America woke up, turned on their mobile phones and logged in to their computers, it was past a million views, and 2,500 comments. By 1 o'clock in the afternoon 'Whispers'[1] had been seen by over 4 million people, retweeted 45,000 times and had over 6,000 comments.

Created by The Lincoln Project, a group of senior disaffected former U.S. Republican Party strategists opposed to Donald Trump's re-election, the Whispers video was one of many that "arrive regularly, whistling in like artillery shells. They are quick, often appearing within a few hours of whatever outrage they are reacting to. They are blunt, not shying away from spotlighting Trump's mental and physical decline".[2]

Whispers seamlessly wove slurs and speculation into 77 seconds of incendiary Information Operations designed to shake the incumbent president's confidence and create fear and uncertainty in his mind. Most of his senior staff and family were featured in the video but one face kept appearing – that of Brad Pascale, Trump's re-election campaign manager and the supposed guru of political marketing on social media. With his distinctive haircut and long beard, Pascale was instantly recognizable. In one shot a black rectangle, more normally associated with obscuring the identity of witnesses or undercover agents, is superimposed on his face but deliberately done without obscuring or anonymizing his features.

---

1 *Trump Whispers*, YouTube, available at https://www.youtube.com/watch?v=9cNpmgeepwI.
2 Steinberg N, *Coalition of the decent' takes on Trump*, The Chicago Sun Times, 7 Jul 2020, available at https://chicago.suntimes.com/opinion/2020/7/7/21315955/lincoln-project-pac-anti-trump-ads-republicans-presidential-election-2020-steinberg.

Pascale was already in trouble with the mercurial president. Just two weeks before he had overseen the disastrous first Trump re-election rally in Tulsa, Oklahoma, when over one million people had allegedly booked tickets, prompting over-flow arrangements in adjacent car parks for the President and Vice President to address the expected huge crowds. Yet on the night the arena was barely one third full. The New York Times reported that "The president, who had been warned aboard Air Force One that the crowds at the arena were smaller than expected, was stunned, and he yelled at aides backstage while looking at the endless rows of empty blue seats in the upper bowl of the stadium".[3] Humiliated, Trump and his staff quickly pointed the finger at Pascale.

Perhaps because of his perceived importance to Trump, Pascale had been no stranger to The Lincoln Project. As the US Spectator magazine wrote, "What Donald Trump hates more than anything is someone making money from his name without cutting him in for a share of the profits".[4] A previous Lincoln Project video had focused on Pascale's multi-million-dollar beach house, his two condominiums, his yacht and his 'gorgeous Ferrari and sleek Ranger Rover'. In fact, US Federal Election Commission filings suggested that Pascale's companies had earned nearly $40 million from various Trump election committees.[5] In the Whispers video Pascale's image first appears just as the narrator tells the audience (the president) that he has "a loyalty problem"; a few seconds later Pascale is again seen, this time peering from behind a curtain as the narrator tells Trump "they expect you to loose".

On 16th July, nine days after the release of Whispers, President Trump announced he was replacing Pascale as his re-election campaign manager.

To those in the business of military Information and Psychological Operations, Whispers was a masterpiece of messaging. Technically it was close to perfection. It had been carefully and precisely targeted to a very specific audience – the President. It had played directly to that audience's

---

[3] Haberman M, *The President's Shock at the Rows of Empty Seats in Tulsa,* The New York Times, 21 June 2020, available at https://www.nytimes.com/2020/06/21/us/politics/trump-tulsa-rally.html.

[4] Wood P, *The trouble with Brad Parscale What if you push the buttons that worked before and nothing happens?* The Spectator, available at https://spectator.us/trouble-brad-parscale-trump-campaign-tulsa/.

[5] Singh M, *Trump replaces campaign manager Brad Parscale in major shake-up,* The Guardian 15 July 2020, available at https://www.theguardian.com/us-news/2020/jul/15/donald-trump-brad-parscale-replaced-campaign-shakeup.

well known and well publicized fears and insecurities;[6]; his absolute demand of loyalty from his staff, his notoriously 'thin skin' and his apparent paranoia. The video had a clear behavioural objective and it had been distributed into a media channel that Trump was known to monitor – Twitter (now 'X'). For additional impact and security, The Lincoln Project had also bought advertising slots in the FOX TV morning news show. These were also well-known Trump favourites. Little was left to chance – the audience would see this video. The Lincoln Project claimed that in just 12 hours "our ad 'Whispers' doubled the average views Tucker Carlson [a known Fox TV talk show host and Trump ally] gets all week". Although it cannot be definitively claimed as being instrumental in Pascale's fall, for as ever the triggers for behaviour are complex, it seems far more than coincidental.

This was not, as Americans might say, a 'Mom and Pop' production. Enormous effort and thought had gone into the video and the various graphics, fades and effects indicate this was professionally edited. Plus the airtime on FOX TV, and other networks, is far from cheap; Federal filings show The Lincoln Project to have deep pockets. In the second quarter of 2020 The Lincoln Project raised nearly US$17 million. Rick Wilson, one of The Lincoln Project's founders told the media that their ads were not just about trolling the president, instead they sought to "litigate the case against Donald Trump",[7] although former Lincoln Project member, John Weaver told Vanity Fair that "driving the president crazy is a short drive". He cited a previous Lincoln Project video which showed Donald Trump at the West Point military academy where he had appeared to have difficulty drinking from a glass of water and then descending a relatively shallow ramp. The Lincoln Project video, 'Something's wrong with Donald Trump' called him out and questioned his cognitive abilities. Weaver told Vanity Fair that: "it [the video] drove him insane and he litigated that for a week and a half culminating his Tulsa toddler activity at that disastrous rally

---

6 For example, see: Post J and Doucette S, *How Donald Trump's Narcissism Masks His Extreme Insecurity*, The Literary Hub 18 Nov 2019, available at https://lithub.com/how-donald-trumps-narcissism-masks-his-extreme-insecurity/; or Graham D, *Trump's Aides Are Desperately Trying to Soothe His Anxieties*, The Atlantic, 12 June 2020, available at https://www.theatlantic.com/ideas/archive/2020/06/soothing-trumps-insecurities-getting-expensive/612985/; or Twenge T, Is Donald Trump Actually Insecure Underneath? Psychology Today, 16 March 2016, available at https://www.psychologytoday.com/gb/blog/the-narcissism-epidemic/201603/is-donald-trump-actually-insecure-underneath.
7 Becker A, *The Group Behind Viral Anti-Trump Ads Wants to Win in November. Its Hyper-Masculine Approach Isn't Helping*, Glamour.com, 15 July 2020, available at https://www.glamour.com/story/lincoln-project-masculine-ads-women-voters.

where for fifteen minute he litigated how he had defeated the ramp like it was the landing at Anzio or something".[8]

On 14th November 2020, on the other side of the Atlantic, I was sent a picture on WhatsApp of the then UK Prime Minister, Boris Johnson with a caption beneath saying "I'd like to thank the people of Liverpool during the mass DNA testing. We expect to solve most crimes within the next few weeks". The caption referred to the establishment, on the 6th November, of a mass COVID-19 testing programme comprising 33 test centres around the city, manned by almost 2000 members of the British Armed Forces. It was a response to what Liverpool Mayor Joe Anderson had called a virus that was now 'out of control' in the city.[9] Anderson would know, his elder brother, Bill, had died of the disease just a few weeks earlier.[10] Whoever had edited the Johnson photo had chosen to replace COVID with DNA and link it to criminality – playing on a long held, and deeply divisive, social stereotype that Liverpudlians were culturally pre-disposed to be 'thieves'. But the potential for mischief ran deeper. The presence of the military in Liverpool, in total around 2000 servicemen and women from 16 units helping with the logistics and delivery of the programme in the city,[11] was a conspiracist's dream.

As visible representatives of the 'deep state' the British Armed Forces were there, in the mind of the conspiracist, not to take blood samples but instead, astonishingly, to insert nano trackers into people's blood stream to allow the deep state to track and control the population. This was already a well-trodden conspiracy path. In May 2020 the Reuters news agency carried a story about social media users sharing a TikTok video that alleged that a microchip would be implanted when the coronavirus vaccine was finally released. Shared thousands of times that video joined the already copious amount of anti-vaccine material online which has dogged health professionals for years. In the UK it had resulted in the World Health Organisation announcing the country was no longer measles free and

---

8 Hagan J, *can negative ads alone beat trump in November?* Vanity Fair, 17 July 2020, available at https://www.vanityfair.com/news/2020/07/can-negative-ads-alone-beat-trump-in-november.
9 Traynor L, *2,000 test positive for Covid since mass testing began and almost 600 had no symptoms*, The Liverpool Echo, 16 Nov 2020, available at https://www.liverpoolecho.co.uk/news/liverpool-news/2000-test-positive-covid-mass-19291271.
10 *Liverpool Mayor Joe Anderson's loss after brother's Covid death*, BBC News, 27 October 2020, available at https://www.bbc.co.uk/news/av/uk-54694740.
11 *COVID: Armed Forces Help Test Almost 20% Of Liverpool Population*, Forces.Net, 13 Nov 2020, available at https://www.forces.net/news/covid-armed-forces-help-test-almost-20-liverpool-population.

declaring that take up of the measles–mumps–rubella (MMR) vaccine had fallen to 87 per cent, lower than the 95 per cent required for herd immunity. UK Prime Minister Boris Jonson was forced to declare that "I am afraid people have just been listening to that superstitious mumbo-jumbo on the internet, all that antivax stuff, and thinking that the MMR vaccine is a bad idea".[12]

As we will see later the most 'successful' conspiracy theories invariably have a grain of truth to them and the TikTok video was a good example in that it had been cropped and taken out of context from a real 2017 NBC News report about a US company which had offered its employees the chance to have a microchip implanted in their finger so they could log in to computers or activate the office photocopying machine.[13]

The 'deep state' conspiracy has existed for years and on both sides of the Atlantic. Supposedly, in the UK, it is a network of behind-the-scenes officials – specifically civil servants and wealthy individuals – who actually run the country, subverting democratically elected politicians. Some believe this to be part of the military-industrial complex, other's see the hand of the Royal Family acting as puppet masters to ensure their continuance. It's been debated in chat rooms, in the nation's newspapers[14] and even during official government enquiries. The Chilcott Inquiry into the Iraq War called former UK diplomat Carne Ross, the UK's representative on Iraq to the UN between 1997 and 2002, to give evidence. Afterwards he told the media: "I testified last week to the Chilcot inquiry. My experience demonstrates an emerging and dangerous problem with the process. This is not so much a problem with Sir John Chilcot and his panel, but rather with the government bureaucracy – Britain's own 'deep state' – that is covering up its mistakes and denying access to critical documents".[15]

In the US the idea of the deep state has really taken root, helped in recent years by repeated references to it by US President Donald Trump who has claimed that the US deep state is the US government itself. Whitehouse

---

12 Burki T, *Vaccine misinformation and Social Media*, The Lancet, October 2019, available at https://www.thelancet.com/journals/landig/article/PIIS2589-7500(19)30136-0/fulltext.
13 *Wisconsin company offers to implant tiny microchips in its employees*, NBC News, 25 July 2017, available at https://www.nbcnews.com/video/wisconsin-company-offers-to-implant-tiny-microchips-in-its-employees-1008326211549.
14 Smith M, *What evidence is there that a 'Deep State' exists in Britain – and is it secretly ruling the country?*, The Mirror, 20 March 2018, available at https://www.mirror.co.uk/news/politics/what-evidence-adeepstateexists-britain-secretly-12012064.
15 Barnet A, *Is there a UK Deep State?* Open Democracy 26 July 2010, available at https://www.opendemocracy.net/en/opendemocracyuk/is-there-uk-deep-state/.

press secretary Sean Spicer told the media that: "I think that there's no question when you have eight years of one party in office that there are people who stay in government — affiliated with, joined — and continue to espouse the agenda of the previous administration, so I don't think it should come to any surprise that there are people that burrowed into government during the eight years of the last administration and may have believed in that agenda and want to continue to seek it."[16] Or in other words, "for the Trump administration and its supporters, the deep state is any part of the apparatus of government itself that doesn't do their absolute bidding".[17]

In the US the deep state story – once mainly the preserve of teenage fans of The X-Files TV series – has taken a more sinister turn with the emergence of the QAnon movement. QAnon is a wide-ranging, conspiracy movement that says that President Trump is having to wage a war against elite Satan-worshipping paedophiles in government, business, and the media. QAnon believers have speculated that this fight will lead to a day of reckoning where prominent people such as former presidential candidate Hillary Clinton will be arrested and executed. Despite Clinton now being out of office she remains a primary target for much of QAnon, so too millionaire philanthropist George Soros and founder of Microsoft, Bill Gates. All of course are from the left of politics; QAnon has little or nothing to say about Libertarian figures such as the Koch brothers. When asked about the QAnon movement and its suggestion that Trump was saving the world from paedophiles and cannibals Trump asked "is that supposed to be a bad thing".[18] In November 2020 Reuters published an article that directly linked QAnon posts to Russia. Citing interviews with current and former Twitter (now 'X') executives and reviewing tweets from suspended known Russian sources, the article concluded that: "From November 2017 on, QAnon was the single most frequent hashtag tweeted by accounts that Twitter (now 'X') has since identified as Russian-backed, a Reuters analysis of the archive shows, with the term used some 17,000 times".[19]

---

16 *Sean Spicer Says White House Believes in Existence of 'Deep State'*, The Huffington Post, available at https://www.huffingtonpost.co.uk/entry/sean-spicer-says-white-house-believes-in-existence-of-deep-state_n_58c30e46e4b0c4575ccb7fbd.
17 Gordon R, *What the American deep state actually is why Trump gets it wrong*, Business Insider, 27 Jan 2020, available at https://www.businessinsider.com/what-deep-state-is-and-why-trump-gets-it-wrong-2020-1.
18 *QAnon: What is it and where did it come from?*, The BBC, 6 Jan 2021, available at https://www.bbc.co.uk/news/53498434.
19 Menn J, *QAnon received earlier boost from Russian accounts on Twitter archives show*, 2 Nov 2020, Reuters, available at https://www.reuters.com/article/us-usa-election-qanon-cyber-idUSKBN27I18I.

In the US the last few years have seen a huge number of conspiracy theories start and spread. The most infamous is probably the ongoing belief that the 9/11 attacks were undertaken by the US government, but others have also commanded international attention. The so called 'Pizzagate' story was promoted by one of the US' most influential conspiracists, Alex Jones, via his Infowars website. Jones, a leading right-wing commentator, claimed that the Hilary Clinton's presidential campaign chairman, John Podesta, had run a pedophile ring in a pizza parlor in Washington DC. Far-fetched? Not is seems to Mr Edgar Maddison Welch who drove nearly 400 miles from his home in North Carolina to shoot his AR-15 semi-automatic weapon at the pizza restaurant. In his defence he told The New York Times that he had gone to rescue abused children – subsequently agreeing that "the intel[ligence] on this wasn't 100%"![20,21]

Alex Jones was also behind the theory that the Sandyhook massacre, a tragedy in which 26 people including 20 children died, was in fact a false flag operation. It appeared that in his mind it was entirely staged, instead, by the US government and so-called victims' families as part of an elaborate plot to confiscate Americans' firearms.[22] Post the 2020 US election supporters of Donald Trump staged a protest rally in Washington DC; US comedian and political commentator Jordan Klepper interviewed a number of attendees for The Daily Show. One man told Klepper that the election had been deliberately compromised by the Coronavirus: "the democrats are working with the communist Chinese to put Biden in because China has blackmail on Joe Biden – that's a fact".[23]

In his book *Conspiracy Theories*, the philosopher Quassim Cassam writes that "it's sometimes suggested that we are living in a 'golden age' of conspiracy theories, but it's actually not clear that conspiracy theories are a hotter topic today than in the past".[24] That may or may not be true but what seems undeniable is that social media has allowed them to spread far wider,

---

20 *Gunman in 'Pizzagate' Shooting Is Sentenced to 4 Years in Prison*, The New York Times, 22 June 2017, available at https://www.nytimes.com/2017/06/22/us/pizzagate-attack-sentence.html.
21 Yuhas A, *'Pizzagate' gunman pleads guilty as conspiracy theorist apologizes over case*, The Guardian, 25 Mar 2017, available at https://www.theguardian.com/us-news/2017/mar/25/comet-ping-pong-alex-jones accessed 15 October 2018.
22 *How Alex Jones and Infowars Helped a Florida Man Torment Sandy Hook Families*, The New York Times, 29 March 2019, available at https://www.nytimes.com/2019/03/29/us/politics/alex-jones-infowars-sandy-hook.html.
23 Jordan Klepper Takes on the Million MAGA March, The Daily Show, 8 Nov 2020, available at https://www.facebook.com/7976226799/videos/889881298486975.
24 Cassam Q, *Conspiracy Theories*, Polity Press, 2021 ed., 32.

quicker and to bigger audiences than was ever the case. As early as 2000 a study in *Journalism & Mass Communication Quarterly* stated, "respondents reported they considered internet information to be as credible as that obtained from television, radio, and magazines …respondents said they rarely verified web-based information".[25]

Former Executive Director of the U.S. Advisory Commission on Public Diplomacy, Matt Armstrong, believes that conspiracies have today been made so much easier by social media which: "allows you to secretly dip your toes into the waters of whatever storyline, with the ability to do many at once. This is why, for example, RT [the Russian English language TV network] run multiple, even contradictory, storylines. Their target audiences, like QAnon, are not critical thinkers"[26]. He calls this the 'marketplace of identity' when you can test drive identities and opinions in ways you could never before, all from the comfort and the secrecy of your own home.

In October 2020 I wrote the Information Operation scenarios for the British Army's Exercise CERBERUS 2020. Around 2400 personnel deployed to one of the UK's largest military exercise areas, Salisbury Plain, to train and validate the army's warfighting planning capabilities. Using the scenario of the Baltic States the nearly four weeklong exercise planned British and NATO responses to an invasion of Latvia by its much larger neighbour Donovia (a pseudonym for Russia). Using the excuse of Donovian (Russian) minorities being persecuted in Latvia, Donovian 'peacekeeping' troops flood across the border, quickly over whelming organic forces. Triggering Article Five of the NATO charter (an attack on one member is an attack on all) NATO forces move quickly to the region. As the scenario develops the criticality of the major bridges over the river Daugava becomes apparent. If Donovian forces can be stopped from crossing, then NATO has a chance of securing the western part of the country. The mission to secure the crossing in the southern city of Jekapils is given to 7th Infantry Brigade British Army (the so called 'Desert Rats'), led in November 2020 by Brigadier Jasper De Quincey Adams.

De Quincey Adams has a very full profile on the business networking site LinkedIn. From that profile anyone can see that as well as a previous operational command (of the Queens' Dragoon Guards), he held the

---

25 Flanagan A and Metzger M, *Perceptions of Internet Information Credibility Journalism & Mass Communication Quarterly* September 2000 77: 515-540, available at https://journals.sagepub.com/doi/abs/10.1177/107769900007700304.
26 Email Correspondence Tatham / Armstrong 2nd March 2021.

important role running the Army's Plans and Policy desk in the MoD's resources and plans department; the incumbents of that assignment sink or swim and if it is the latter then they are destined for higher rank. His profile also shows that he was Colonel Team leader in Task Force Ukraine before moving on to spend 2 ½ years as the Special Advisor to the Chairman of the NATO military committee in NATO headquarters in Brussels. Critically, De Quincey Adams' posting to Ukraine coincided with the loss of Malaysia Airlines Flight 17 (MH17), a scheduled passenger flight from Amsterdam to Kuala Lumpur, shot down on the 17th July 2014 while flying over eastern Ukraine. All 283 passengers and 15 crew were killed.

In May 2018 the Dutch led investigation team concluded that MH17 had been shot down by a missile belonging to the 523rd Anti-Aircraft Missile Brigade (Russian Army) based in Kursk. Dutch Foreign Minister Stef Blok told the media that: "On the basis of the [joint international team's] conclusions, the Netherlands and Australia are now convinced that Russia is responsible for the deployment of the Buk installation that was used to down MH17".[27] Throughout that investigation Russia had denied all responsibility. In July 2015 it had vetoed a draft UN Security Council resolution to set up an international tribunal, claiming it would be premature. Subsequently the Russian Foreign Ministry called the joint Dutch and Australian findings "biased and politically motivated". Russia has meanwhile advanced numerous theories about the loss of the aircraft, many of them revolving around Ukraine forces (in league with NATO) shooting down the aircraft having mistaken it for President Putin's presidential jet.

In Exercise Cerberus 2020, Information Operations would be used by both sides to try and achieve military effect. The race to the river crossing by 7th Infantry Brigade and the Brigadier's detailed LinkedIn account was too good an opportunity not to exploit and a totally fake Twitter (now 'X') persona was invented ('Janis Berzina'). 'Janis' was a Donovian (Russian) speaking Latvian and union activist who earlier in the exercise had called for some low-level civil disobedience against the Latvian government. As a result, his online presence had been built up and he had (for exercise purposes) become an important voice and influencer in the ethnic Donovian diaspora. As 7th Infantry began their move towards the bridge, Berzina began to tweet about how they were led by a British military officer – De

---

27 *MH17 Ukraine plane crash: What we know*, BBC News, 26 Feb 2020, available at https://www.bbc.co.uk/news/world-europe-28357880.

Quincey Adams – who had been advising the hated Ukrainian governed when MH17 was shot down. As a Donovian (Russian) supporter Berzina bought into the Russian narrative completely and De Quincey Adams was fictitiously portrayed as a British 'war criminal' who had been instrumental in advising the Ukrainian government into the shooting down of the jet. And because the British were led by a 'criminal', Berzina would do all he could to stop De Quincey Adams from committing ethnic cleansing of the native Donovian population. Berzina called for mass protests and a rally on the bridge to stop 7th Infantry Brigade from securing it. Of course all of this was a complete lie, a deception, based upon just a small grain of truth. De Quincey Adams was a distinguished British Officer who had enjoyed an impressive and faultless military career. But, for our adversaries, exercise or real, the truth would be irrelevant and this exercise was designed to simulate exactly what our adversaries would do in a real conflict, creating a level of unwelcome complexity that would tie up the planning staff and, perhaps, slow down the advance on the ground. Anyone with a public profile – and particularly senior leaders such as De Quincey Adams – can be exploited as part of disinformation campaigns. The point of Exercise Cerberus was to demonstrate just that – even if it was to the discomfort of the blameless and entirely innocent Brigadier – and to test serving commanders on what they would do to combat aggressive disinformation.

Such exercises are vital. As we see in the next chapter the military have a long history of Information Operations but surprisingly has seemed remarkable slow in adapting to the post social media world. Take Exercise JADE HELM 2015, a US Special Operations Command sponsored exercise to improve their unconventional warfare capability, conducted throughout Texas, New Mexico, California, Nevada, Utah, and Colorado. One of many hundreds of military exercises conducted each year by US forces around the globe. Jade Helm became infamous for its 'hijacking' by, again, Alex Jones. Jones told his many followers that Jade Helm was actually a plot by the Obama administration to implement sharia law in the United States and would involve the confiscation of citizens' weapons. As utterly ridiculous as its sounds 32% of republicans told a public polling survey that they thought the Federal Government was trying to take over Texas.[28] Both of these examples had a factual truthful genesis but today it is just as likely

---

[28] Walker Leads Tightly Clustered GOP Field, Clinton Up Big Nationally, Public Policy Polling, 13 May 2015, available at https://www.publicpolicypolling.com/wp-content/uploads/2017/09/PPP_Release_National_51315.pdf.

that issues will be manufactured to suit an agenda. This has never been truer than in Ukraine, which perhaps more than any other single conflict in history has perhaps presented the most significant information challenges.

Quite literally there are thousands of examples of information being used to manipulate and distort but the following examples illustrate both ends of a successful spectrum – simple image manipulation through to complex multi-dimensional attacks. At the simple end of the spectrum is basic image manipulation – something that anyone with a passing knowledge of software such as Adobe Photoshop, would easily be able to replicate. And they do. @GogiGogi12 is a Twitter (now 'X') user with over 2,000 followers. It describes itself as a 'proud gremlin from the kremlin'. In 2014 @GogiGogi12 tweeted a picture of a little girl, covered in mud, forlornly holding her puppy. The title of the tweet was 'UE-US criminals – Save Donbass people'. The image was designed to trigger an emotional response, suggesting that the Ukrainian (UE) government – supported by the US – had been responsible for the little girl's plight. But all was not as it seemed. The picture of the little girl with her puppy, covered in mud, was actually taken in Australia and had been entered into a photographic competition run by the Northern Territories Cattleman's Association, entitled MuddyLove. Its photographer, Caterine Atkins, had won first prize.[29]

Another classic example of basic image manipulation appeared on 27th September 2011 and featured Alexander Skryabin, who had died of lung cancer. Which makes his decision to be a suicide bomber, blowing up a Ukrainian tank with a bundle of grenades on 7th July 2014, three years after his death, quite improbable. Regardless, Skryabin's photo appeared in a series of articles, many of them declaring him to be a hero of Novorossiya. Some days later photos of Skyrabin's funeral appeared online, again eulogising him for his action. These latter funeral photographs were in fact taken from the burial of Ukrainian miners taken in August 2011. No one seemed too perturbed that the story was fake, although highly unusually the Russian TV channel RT did subsequently issue a retraction: "the information presented in it did not correspond to reality: Alexander Scriabin, who is referred to in the material, died in 2011".[30]

In Spring 2015 the Swedish Security Service, Säpo, stated in its annual report that Russia's intelligence activity was the greatest threat to

---

[29] McPhee L, *Puppy and mud a winning combination*, West Australian, 1 April 2020, available at https://thewest.com.au/news/australia/puppy-and-mud-a-winning-combination-ng-ya-217042.
[30] Available at https://russian.rt.com/article/39825.

Swedish security – and that every third Russian diplomat in Sweden is a spy.[31] As to prove the point, later that year Twitter (now 'X') circulated what was portrayed as a leaked letter from the Swedish Defence Minister, Peter Halqvist, to a Director of British Defence company BAe Systems, Magnus Igesso, thanking him for showing off the archer missile system to Ukrainian colleagues.[32] For formally neutral Sweden this was a big issue; sowing seeds of discontent within its population would be helpful to Russia, keen to ensure Sweden's continued neutrality. Halqvist issued an absolute statement that the letter was fake; by whom remains unknown.

In all these examples we see information being used to manipulate and persuade audiences. In 'Whispers' it's in the political context with a defined political objective; in Liverpool the objective is less clear and may be as simple as mischievousness or something more sinister playing as it does to offensive and patently untrue stereotypes ('all scousers[33] are thieves') and in Exercise CERBERUS it is a clear military objective. But all three are now being conflated and invariably referred to, often pejoratively, as Information or Psychological Operations – two military terms that have been utterly subordinated to describe what are in many examples simply lies and deliberate disinformation for personal, financial, or political gain. That conflation is in part based on understandable ignorance by a general public that has no specific understanding, sees the terms and thinks they fit. But in part it is being driven by elements of the media and, as we will see later, by certain academic voices who in recent years have embarked on a deliberate path to discredit and demonise not just IO as a vital military discipline but so too its practitioners.

So just what exactly are military Information Operations? If we believe what is written online and in parts of the media, then military Information Operations are part of the deep state, are mind altering villainous activities responsible for Brexit, for the election of Donald Trump, for the hoodwinking of media, the public, civil society, and the death of democracy. Information Operations are perpetuated by unprincipled and unethical practitioners, many in the pay of shady political or commercial organizations. Information

---

31 Lapidus A, *Härifrån jobbar Putins spioner i Sverige*, Expressen.se, 20 Sept 2015, available at http://www.expressen.se/nyheter/harifran-jobbar-putins-spioner-i-sverige/.
32 Sweden offers to sell weapons to Ukraine in the forged letter available at http://translate.google.co.uk/translate?hl=en&sl=sv&u=http://www.svd.se/nyheter/inrikes/forfalskat-brevpastar-att-sverige-ska-salja-vapen-till ukraina_4353755.svd&prev=search.
33 A colloquial term for residents of the city of Liverpool in North West England.

Operations should be outlawed; their practitioners should 'be ashamed'[34]; they should be monitored, regulated. Information Operations, and its corollary of Psychological Operations, have such powerful Orwellian associations they have become irresistible to conspiracists and critics globally who are quick to point to their hand in, well, everything! This has not been helped by one of the biggest global news stories of the decade, the Cambridge Analytica affair, or that one of the maddest and baddest internet conspiracists – US radio show host Alex Jones who calls his site InfoWars and see's false flags, conspiracies and so-called 'blacks ops' behind every possible event. This narrative has proven very compelling and, in many instances, challenging for military Information Operations. As we will see later some nations have pulled back from their Information Operations capabilities as a result – and at the exact moment that adversaries such as Russia, Iran and China have stepped up and even accelerated theirs.

The trouble with the 'popular' definition of Information Operations is that it's simply untrue but to uncover this requires not just a dive into history but so too into military doctrine. Doctrine is the fundamental set of principles that guide military forces; it is designed to provide the bedrock of all military thinking and planning. The trouble with doctrine is its dry, often quite ethereal and through the years has become increasingly laboured in its presentation. It is certainly not an easy read and prompted one ex British Army Officer, and latterly Professor of War Studies, Jim Storr, to write that "the British Army used to value simple, clear English ... What we see now is often garbage. I have critiqued several drafts of British army doctrine over the last few years and (privately) ridiculed the language used[35]"

Whilst the latest doctrine will always feature on every young military officer's reading list, at least presentationally, it's probably debateable how much actually gets read by soldiers in the field. Indeed, there is an amusing anecdotal quote from an unnamed Soviet junior officer, that has for years done the rounds of military staff colleges, that: "One of the serious problems in planning the fight against American doctrine, is that the Americans do not read their manuals, nor do they feel any obligation to follow their doctrine".

NATO, an alliance of 30 nations, like most military organisations, has copious doctrine. The trouble with NATO doctrine is that unlike that of

---

34 A comment directly levelled at the author via social media.
35 Storr J, LinkedIn Post 4 January 2024, available at https://www.linkedin.com/in/jim-storr-1209962/recent-activity/all/.

the US or the UK, it has to be approved by all 31 member nations and so the process of its creation is both tortuous but reassuringly rigorous; when NATO settles on a definition or a concept you can be sure it has really been thought through. NATO defines Information Operations as:

> "coordinated and synchronized actions to create desired effects on the will, understanding and capability of adversaries, potential adversaries and other approved parties in support of the Alliance overall objectives by affecting their information, information-based processes and systems while exploiting and protecting one's own."[36]

Wow! Neither short, nor interesting nor punchy. And one of a number of terms that have through the years seemingly become interchangeable – even amongst the military (and thus perhaps proving the junior Soviet staff officer, above, right), who don't appear to have read their own doctrine. Terms such as Propaganda, Information Warfare (IW), Psychological Operations (PsyOps in UK and NATO and PSYOP in US doctrine), Strategic Communication (StratCom), Information Activities (IA), Information Exploitation (IE) and Influence. Confused? You would not be alone. Even the US Department of Defence – the Pentagon – freely admits that its confused by the various terms. To help it, it commissioned its own review on 'information warfare' and all the related terms and concluded that: "even within the Department of Defense, the terms have had elastic, imprecise and ambiguous meaning and are often used interchangeably to describe activities that are divergent in nature".[37]

We need some help. Back in the 1960s an engineer at the US defence company Lockheed Martin's advance development program (known as the Skunk Works) told his team of aeronautical engineers that whatever they designed it had to be fixable by the most average mechanic in the most arduous combat conductions. That direction became known as the KISS principle – KISS being an acronym for 'Keep it Simple Stupid' – a term that is now widely used across western armed forces. So, keeping it simple it's worth having a quick discussion about each of the terms and then settling on one for the rest of the book.

---

36 Allied Joint Doctrine for Psychological Operations (AJP3-10), available at https://www.gov.uk/government/publications/ajp-3101-allied-joint-doctrine-for-psychological-operations.
37 Lin H, *Doctrinal Confusion and Cultural Dysfunction in the Pentagon Over Information and Cyber Operations*, LawFare, 27 March 2020, available at https://www.lawfareblog.com/doctrinal-confusion-and-cultural-dysfunction-pentagon-over-information-and-cyber-operations.

Propaganda

Let's start with Propaganda. This historic term is, linguistically, probably the most accurate of them all to define what it is that Information Operations people 'do'. The trouble is that it has through time also become one of the most tarnished which is why today it is almost invisible from the formal doctrinal lexicon of the UK, US, and NATO. The term itself originated in Europe in 1622, shortly after the start of the Thirty Years' War, which pitted Catholics against Protestants. Pope Gregory XV founded the Sacred Congregation for the Propagation of the Faith ('*sacra congregatio christiano nomini propaganda*') a Vatican department charged with the spread of Catholicism; with countering Protestantism (the church had been particularly stung by Martin Luther's[38] cartoons of the 'Donkey-Pope of Rome'[39] which appeared in a series of pamphlets and books and which were emulated by other protestant reformers such John Knox and John Calvin); and with the regulation of ecclesiastical affairs in non-Catholic countries.

Originally the term was not intended to refer to misleading information, in fact it is closely related to the word 'propagate' which we more commonly think of in terms of growing new shoots, and in the context of military Information Operations it seems entirely apt since much of the work is associated with encouraging people to think again. In 1940 John Hargrave published 'Propaganda. Words Win Wars' and in defining propaganda he wrote, 'there is no need to be scared of the word propaganda, it is a perfectly good Latin term .... In the popular mind however it has come to mean something Jesuitical, something Machiavellian, a system of organised falsehood, cunningly devised, inherently evil – a 'lies factory – and therefore to be shunned by honest men'.[40]

And shunned it has been, largely because of misuse by such infamous individuals as Nazi Minister Joseph Goebbels and Soviet Leader Joseph Stalin, who both saw propaganda as a means of indoctrinating their respective populations, and ostracizing components of it. And so today the Oxford English Dictionary defines propaganda as information, especially of a biased or misleading nature, used to promote a political cause or point of view.

---
[38] Martin Luther was a German priest and academic and one of the instigators of the 16th Century reformation in Europe which led to the split between Catholicism and Protestants.
[39] A picture of one of them can be see here: https://www.alamy.com/stock-photo-caricature-of-the-pope-by-lucas-cranach-the-elder-with-text-by-martin-131473426.html.
[40] Hargrave J, *Words Win Wars*. Propaganda The Mightiest Weapon of all', Well Gardner Darton & Co, 1940.

Information Warfare

The next term is Information Warfare, again now routinely used in the media and by the public but which in the military context tends to define a range of activities that occur in the electromagnetic spectrum – for example denying the opposition the use of specific radio frequencies and bandwidth through jamming or degrading IT systems. In military terms this is more properly known as CEMA (Cyber and Electromagnetic Activities). The defence company BAE Systems has a very helpful if slightly jargonistic online guide to CEMA.[41] CEMA is not however widely understood in the public domain and perhaps because of its complexity the term Information Warfare is commonly used. This term now embraces the rapidly emerging military interest in cyber – be it protection of data (defensive cyber operations) or the much sexier (and less discussed) offensive cyber operations against an adversary. As we will see in a later chapter with the example of the Russian hacking of Kyiv's traffic lights, when cyber and Information Operations are sewn together they can become a very sophisticated weapon.

Psychological Operations

Psychological Operations – abbreviated to PsyOps in the UK and PSYOP in the US, is perhaps one of the most publicly misunderstood and abused terms. In NATO and most western military doctrine, PsyOps is 'truthful and attributable' information that targets an adversaries' attitudes and behaviours; the key words here are 'adversary' – as in military threat and being 'truthful and attributable'. This is not what many commentators will have you believe; to many PsyOps is seen as some kind of mind control and brain washing directed across domestic populations and the idea of truth is very far away from their imagined definition. The term is however Orwellian and unhelpful and in 1999 the UK sought to render it obsolete when 15 (UK) PsyOps Group (15 POG) was renamed 15 (UK) Information Support Group. As 15 PsyOps files reveal, rather bizarrely, British Armed Forces corporately then forgot it had a PsyOps capability and demand for PsyOps support fell whilst requests for IT support dramatically increased! When the US changed PsyOps to MISO (Military Information Support to Operations) in 2010 the UK perhaps wisely decided not to follow. Regardless

---

41 BAE Systems, *Cyber and Electromagnetic Activities (CEMA) Integration*, available at https://www.baesystems.com/en/digital/solutions/defence/cema-integration.

of name, PsyOps capabilities exist across most western armed forces – and in sizeable numbers in the US, in Germany, Italy and Romania.

Strategic Communication

Strategic Communication is a comparatively new term and is primarily a philosophy, partly a capability and partly a process. The UK defines it as: 'the systematic and coordinated use of all means of communication to deliver UK national security objectives by influencing the attitudes and behaviours of individuals groups and states', an unhelpful definition since it includes the word 'influence' which has in turn now entered the military lexicon under its own steam. When used as a noun the military see it as an outcome but when used as a verb it is in the context of specific activities, many of them planned and coordinated by IO staff and embedded, covertly or overtly into wider strategic communication plans.

Information Activities

One of the newest of all the terms – Information Activities[42] – is now increasingly a 'catch all' for all of the above and, unmentioned thus far, the military's interaction with the media which in the US is called Public Affairs (with its constitutional constraints – to inform, not influence, US audiences) and, in the UK, Media Operations. Also included within Information Activities is Military Deception (MilDec) which includes the use of camouflage and concealment to hide a force, the use of feints and ruses to get an enemy commander to make bad decisions and deception in support of Operational Security (OPSEC) to mask a friendly commander's intent – all of which is conducted to protect the lives of friendly forces. Operation Mincemeat, conducted during World War II to disguise from the Germans the invasion of Sicily in 1943, perhaps being the most famous example.

Misinformation and Disinformation

And finally, two more terms that are often used interchangeably but which are in fact quite different: misinformation and disinformation.

---

42 Pomerleau M, *Out: 'information warfare.' In: 'information advantage'*, C4ISRNET, 29 Sept 2020, available at https://www.c4isrnet.com/information-warfare/2020/09/29/out-information-warfare-in-information-advantage/.

Misinformation, which according to some online references was used in the late 1500s,[43] combines the word *information* with the prefix *mis* – meaning 'wrong' or 'mistaken'. Or put another way, *misinformation* doesn't care about intent, it is simply a term for any kind of wrong or false information. In the US misinformation is defined as 'the spreading of unintentionally false information. Examples include internet trolls who spread unfounded conspiracy theories or web hoaxes through social media, believing them to be true'.[44] Disinformation is also wrong information but in this case, it is information that is deliberately spread. Compared to *misinformation*, *disinformation* is a relatively new word, first recorded in 1965–70 and it's said to derive from the Russian word *dezinformatsiya*. The same US Congressional briefing document provides some examples: 'Examples include planting false news stories in the media and tampering with private and/or classified communications before their widespread release'.[45]

It's a lot to take in so, for the purpose of this book, lets agree on a single term that we will use throughout. I'm going to use Information Operations as a generic term to mean everything that gets done in the information environment but with one important exception – cyber. I'm going to use cyber as the term for offensive hacking and defensive protection of computer systems. When I talk about military Information Operations (at least western ones) I am referring to very tightly controlled processes – all defined in doctrine and many of them legally defined in the Law of Armed Conflict and Rules of Engagement – designed to inform and influence for effect. And critically, they are proceses and tools that are used against adversarial forces, and their political leadership, in conflict environments, not home domestic populations.

Accurate and attributable? If you believe the internet, and certain academics, then Information Operations is about deceit and fake news. And that, sadly, is what many people think. And again the water is muddied because Information Operations will be used to support military deception operations – and that may very well involve being less than truthful and transparent, because to do so would risk the lives of those deploying on

---

[43] *"Misinformation" vs. "Disinformation": Get Informed on the Difference*, dictionary.com, 15 August 2022, available at https://www.dictionary.com/e/misinformation-vs-disinformation-get-informed-on-the-difference/.
[44] Theohary C, *Defense Primer: Operations in the Information Environment*, US Congressional research Service briefing document dated 14 December 2023. Available at https://crsreports.congress.gov/product/details?prodcode=IF10771.
[45] Ibid.

What Are Information Operations? 29

the UK's behalf. The example of Sierra Leone, in the foreword, was a prime example. The fire power demonstrations were in a sense misleading, in that the UK's Rules of Engagement prevented offensive operations against the RUF, only defensive were allowed. But the RUF didn't know that and the clear intent of the landings and fire power demonstration was to create fear amongst the RUF. The next chapter begins with another lie – the so called 'Eastern Front Eunuchs' – and again makes people believe that Information Operations are simply 'fake news'.

Like everything else in the military, deception – which is known specifically as MILDEC (military deception – deception by the military, against an enemy force for military objectives) is carefully described in doctrine which you can find online[46]. The US defines it as "those actions executed to deliberately mislead adversary decision makers as to friendly military capabilities, intentions, and operations"[47] and a little later on it explains that commanders can "use any of their forces and all available methods subject to the rules of engagement and law of armed conflict to accomplish their MILDEC objective".[48] But deception is not Information Operations; Information Operations simply helps through good OPSEC (Operational Security) (there is no point attempting deception if the enemy already knows your intent) and appropriate Presence-Posture-Profile (PPP – the force's body language). For deception to work it needs Information Operations, but this does not mean that Information Operations is inherently deceptive. For example, Media Operations – the UK military term for Public Affairs – would never deliberately give journalists inaccurate information in an attempt to deceive – but would also not give them the entire plan. A good Information Operations staff officer would identify those accurate and truthful pieces of the plan that may support a deception and allow the intended audience to draw their own conclusion. The best example of this is the famous Left Hook in the 1991 Gulf War. Saddam Hussein did not trust his intelligence officers and watched CNN, he also believed the Coalition would liberate Kuwait from the sea. So the Coalition deception included embedding CNN with the US Marines in the Gulf and noisy Navy SEAL missions onto the beaches around Kuwait City. All to reinforce an existing perception, not to lie to the media.

---

[46] For example, see *Allied Joint Doctrine for Operations Security and Deception (AJP-3.10.2)*, available at https://www.gov.uk/government/publications/allied-joint-doctrine-for-operations-security-and-deception-ajp-3102a.
[47] *Joint Publication 3-13-4 Military Deception*, US DOD, 14 Feb 2017, available at https://fas.org/irp/doddir/dod/jp3_13_4.pdf.
[48] Ibid.

In our earlier examples, the opeations in Sierra Leone were against an accepted military adversary, undertaken with the agreement of a democratically elected host government led by President Kabbah of Sierra Leone, and were very carefully controlled and legally tested. In the example in the next chapter, it was part of a war of national survival. In Sierra Leone the deception revolved around the significant and real physical presence of military troops, and it was successful. Truth is important, but never at the expense of compromising an operation and putting lives at risk. The military has learnt, through bitter experience, that falsehoods and lies will almost invariably always be found out and that it actually yields far less successful results than the truth. Both examples were against a recognised military adversary and never against a domestic population. The issue of the military not targeting domestic audiences is really important. In the US, for example, this is strictly regulated by a series of complex rules and laws. If ever broken, which regrettably has occasionally happened, there are legal consequences.[49]

The last bit of the Information Operations definition, which often gets forgotten is 'protecting one's own'. It has always been important to protect information but arguably never more so when not only can information be globally distributed within seconds, but soldiers, sailors and airmen are as connected to social media as their civilian colleagues. Time and time again we see OPSEC breaches. Unsurprising the Russian parliament banned its soldiers from using smartphones whilst on duty after numerous lapses,[50] particularly during the MH17 downed airline incident when social media investigation companies such as Bellingcat were able to track the movements of the specific Russian missile launcher accused of shooting down the aircraft via social media postings.[51] However, as we will see later, personal lapses in OPSEC have proven costly for both sides in the Ukraine / Russian conflict. As former US army intelligence officer Zachery Tyson explained on Twitter (now 'X'): "in military terms information acts like a biological weapon. Like a virus it can infect the host and spread".[52]

---

49 Mulrine A, *US Army may have used PsyOps against Senators. How is that different from PR?*, The Christian Science Monitor, Feb 2021, available at https://www.csmonitor.com/USA/Military/2011/0225/US-Army-may-have-used-PSYOP-against-senators.-How-is-that-different-from-PR.
50 *Russia bans smartphones for soldiers over social media fears*, BBC News, 20 Feb 2019, available at https://www.bbc.co.uk/news/world-europe-47302938.
51 Gibney J, *How social media cracked the case of MH17*, M Today, 19 Oct 2015, available at https://www.todayonline.com/innovation/how-social-media-cracked-case-mh17.
52 Twitter (now 'X') @ZacheryTyson.

Understanding its effects in groups and audiences, and applying critical thinking, might perhaps be viewed as a 'vaccine' and Information Operations is the vaccine's 'jab'.

And finally! Typing the words *Information* and *Operations* (and indeed reading it) hundreds of times is going to be pretty tedious so from now on I will just use the term IO.

2 Britain and IO
From Eunuchs to the Archers

In the basement archives of the Ministry of Defence (MoD) in London, and in the MoD's Top Secret file archive in HM Naval Base in Portsmouth, shelves of now dusty and long unopened files chart the UK's long flirtatious relationship with the use of information to achieve effect – variously being described as propaganda, Psychological Operations, Information Operations (IO) and, today, Information Warfare and perhaps more benignly, Influence. Through the years those organisations charged with managing that relationship have been involved in some of the major conflicts of the twentieth and twenty-first century: World War I, World War II, the 1953 Suez Crisis, Northern Ireland, Bosnia, Sierra Leone, Iraq, Afghanistan, Libya, Syria and Somalia. Yet these have merely been the most publicly visible tip of a larger iceberg. In British Honduras, in Hong Kong, Indonesia, Jamaica, Malta, Borneo and countless other areas of the globe, long forgotten operations and skirmishes have seen HMG deploying information to attempt to positively influence outcomes.[1] IO is far from new.

Surprisingly very few serving UK military personnel today will know much about these operations at all. To progress the ranks of the Army, Royal Navy or Royal Air Force, officers are expected to become masters of infantry tactics, engineering, logistics, and address threats in terms of manoeuvring personnel, tanks, ships, and aircraft. It is only in recent years that IO has slowly begun to be recognised as an important capability in its own right but even today, post Ukraine, for some it remains a peripheral activity – 'sprinkled on' as an afterthought in the military planning process. During the operations in Iraq and Afghanistan, which in the military are referred

---

1 These operations are covered extensively in the literature. For example see Carruthers, *Winning Hearts and Minds. British Governments, the Media and Colonial Counter-Insurgency 1944-1960*; Vaughan *'Cloak Without Dagger: How the Information Research Department Fought Britain's Cold War in the Middle East 1948-56*; Smith, *General Templer and Counter-Insurgency in Malaya: Hearts and Minds, Intelligence and Propaganda*, Intelligence and National Security, vol 16 no 3 (Autumn 2001); Dyer, *The Weapon on the Wall – Rethinking Psychological Operations Warfare*; Welch, *Propaganda and Persuasion* and Lashmar and Oliver's *Britain's Secret Propaganda War 1948-1977*.

to as COIN (counter insurgency) several important military thinkers began to talk about the centrality of information. Dr David Kilcullen, a former Australian Army Officer, wrote that: "We typically design physical operations first, then craft supporting information operations to explain our actions. This is the reverse of al-Qaida's approach. For all our professionalism, compared to the enemy's, our public information is an afterthought. In military terms, for al-Qaida the 'main effort' is information; for us, information is a 'supporting effort.'"[2] This was absolutely right but as those that served in the deserts of the Middle East and South Asia diminish so the lessons fade.

Measuring the effectiveness of IO is never easy but the evidence suggests that there have been successes – so too, failures. Indeed the failures are probably more instructive than the successes. What is indisputable is that through the years a small group of military officers and civil servants have quietly worked to mitigate and reduce risk to the UK, and to its many allies around the world, using information as their weapon of choice. Paradoxically that supposed expertise is widely internationally recognised. Be it in direct support of international allies or in a training and mentoring role, the UK is widely, if perhaps slightly undeservedly, courted and seen to be expert in this specialised area. In large part that still remains a legacy of operations from World War II, when alongside the Churchill's Special Operations Executive (SOE) – established to 'set Europe ablaze'[3] – many other less well-known units were quietly set up to demoralise, influence, persuade and harry the morale of Germany.

In July 1940, just weeks after the German conquest of France and the Low Countries, Adolf Hitler decided to attack the Soviet Union and on 18th December 1940 signed Directive 21[4] (code-named Operation 'BARBAROSSA'), authorising the invasion of the Soviet Union. The permanent elimination of a Communist threat to Germany, and the acquisition of land to the east for 'lebensraum' (living room) had been a core policy and ideological driver of the Nazi movement since the 1920s. That he chose 1941 to do so was ultimately to change the course of World War II, allowing the UK, in particular, to recover from the disasters of Dunkirk,

---

2 David Kilcullen, *Countering the Terrorist Mentality*, New Paradigms for 21st Century Conflict available at https://smallwarsjournal.com/blog/new-paradigms-for-21st-century-conflict.
3 *The Secret British Organisation of the Second World War*, The Imperial War Museum, available at https://www.iwm.org.uk/history/the-secret-british-organisation-of-the-second-world-war.
4 *Invasion of the Soviet Union, June 1941*, United States Holocaust Museum, available at https://encyclopedia.ushmm.org/content/en/article/invasion-of-the-soviet-union-june-1941.

reform, re-equip and retrain its armies and, with the assistance of huge numbers of free European forces, and the entry of the US into the war after the attack on Pearl Harbour, ultimately return to Europe via the beaches of Normandy three years later.

For the German Army it was to mark the beginning of the end and the capitulation of the German 6th Army in February 1943 after the battle of Stalingrad was decisive. For the Soviet Union it marked the start of the most horrific war of annihilation in which over 20 million Russian civilian and military personnel would die. Although the western allies were not in a good position militarily they sought to offer what [limited] assistance they could, principally via the shipping of supplies via Atlantic convoys into the northern Russian port of Murmansk – an operation that battled not just Nazi forces but the hardships of the North and Baltic Seas and the brutality of the Russian winter,[5] a winter for which the German troops were singular unprepared and which spawned the rumour of the Eastern Front Eunuchs.

Whilst Nazi troops initially made good progress, forward units even coming in sight of Moscow, they were quite literally stopped in their tracks by the onset of winter. Heavy autumn rains turned roads to mud and dramatically slowed not just progress of forward units but the long supply chains of fuel, ammunition, and food necessary to keep an army advancing. But the rains were followed by the onset of winter, with temperatures dropping as low as 40 degrees, freezing tanks and equipment, shutting down diesel engines and having a disastrous effect on German soldiers who had not been equipped with winter provisions by a leadership utterly convinced in its ability to crush the Russian nation by the end of the Summer. Missing coats, hats, proper boots and gloves, soldiers froze to death on the exposed and wind blasted Russia steppes.

In London, in the offices of the newly established Political Warfare Executive (PWE), opportunities to exploit this situation were examined and resulted in, amongst others, proposal R695 being tabled on 7th November 1941.[6] The idea was for Special Operations Executive (SOE) agents in Portugal, Spain and Tangiers to brief their contacts in the local media and 'gossiping networks' that German soldiers on the eastern front were having their genitals amputated because of severe frostbite. The idea was fostered

---

5 An operation that this author's Grandfather was part of, serving onboard the Royal Navy warship HMS Pozarica.
6 R/695 National Records Office available at https://www.psywar.org/content/sibNetSearch.

by an intercepted signal in October 1941, from a German Medical Corps commander, that amputations across the front as a result of frost bite, had trebled in the last two weeks.[7] The original signal made no reference to genitals but the PWE decided that this was a worthwhile rumour to spread amongst neutral press in the hope of it being picked up in occupied nations and, most particularly, in Germany. This would, it was felt, undermine Nazi claims of success on the Eastern Front and, it was hoped, directly target the morale of the wives, mothers and sisters left behind in Germany and undermine their confidence in the Nazi leadership.

And so began the legend of the Eastern Front Eunuchs. How widespread that rumour became and what success it enjoyed, if any, is unknown. The Public records office files for the PWE are missing huge swaths of data for 1942 (due to fire damage) and nothing is known to be recorded in German war archives. However, it is understood that the story appeared in at least one US newspaper outlet.

By 1942 the PWE had some 458 propagandists working for it, most based at Woburn (a short drive from Bletchley). Amongst its many activities were the creation of fake radio stations. 'Gustav Sigfried Eins' (GSE) – which alongside 'Soldatensender Calais' (SC) – was the brainchild of one of PWE's most famous figures, Sefton Delmer, a former Daily Express newspaper journalist, fluent German speaker, personal acquaintance of many of the most senior Nazi leaders and, as his memoires reveal,[8] eccentric patriot with a great eye for innovation and opportunity. Broadcast from a 500kw medium wave transmitter known as Aspidistra (named after a music hall song made famous by Gracie Fields) GSE and SC purported to be German radio stations and they were to prove adept at spreading confusion and sedition.[9] Soldatensender Calais had the distinction of being the first radio station to broadcast 'news' of the D-Day invasion to German forces.

Another of Selmer's black projects was a station he called 'Christ the King' which aired 'talks' provided by a 'Father Andreas', who denounced the Nazi attacks on the church and the dislocation of Christian family values for Nazi ones. Delmer, in his autobiography, stated that he actively sought to spread rumours (sibs – from the latin word *sibiliare* meaning whisper)

---

7 R/669 National Records Office. Available at https://www.psywar.org/content/sibNetSearch.
8 Delmer S, *Black Boomerang* (Secker & Warburg, 1962); available to download at www.seftondelmer.co.uk.
9 It is fascinating that the history of the Political Warfare Executive was hidden from view until the publication of *The Secret History of the PWE* by David Garnett, some fifty years after it was written.

around Europe that the station was actually run not out of the UK, but the Vatican – a course of friction to the Vatican who until as late as 1978 were still publicly denying their involvement.[10]

One of the foremost authorities on the PWE is British historian, Lee Richards, who in his two excellent books, *'Whispers of War'* and *'The Black Art'* lists some of the many other sibs created. Lee tells us of a rumour that RAF planes had the capability to set the channel alight in the event of an attempted German invasion[11,12]; the faulty construction of German U-Boats; that SS soldiers were being stationed in safe towns in Germany where they were taking advantage of the wives and girlfriends of Wehrmacht soldiers fighting on the front lines; that German doctors had been instructed to let elderly patients die in order to ease rationing problems. Whilst all of these were patently untrue, all had the veneer of truth. As we will see later, even the smallest 'grain' of truth can allow the most extraordinary claims to appear authentic. Take the issue of setting fire to the channel. The British Army had discarded Flame Throwers from their order of battle (ORBAT) after World War I. As the German army thundered across Europe, the British Ministry of Defence set about reconstituting the capability and conducted some trials in St Margaret's Bay on the UK's south coast which resulted in small areas of pre-prepared water catching fire for a short period of time. Concurrently, and fortuitously, two further events occurred that would give the story 'legs'.

The first was a routine RAF bombing raid over Calais which accidentally caught a German battalion on invasion exercises and a large number of personnel suffered very bad burns. The second was the capture, by the Royal Navy, of 10 German soldiers, posted as guards on French fishing trawlers. The guards – probably downgraded from their normal duties due to illness or injury – had been drawn from eight different units. The PWE run radio stations now had a series of 'facts' on which to propagate their sib. The radio named the soldiers and their units, telling listeners that they had been rescued but that the fate of most of their comrades was unknown although some were being treated for bad burns. For added drama Wagner's funeral march was played. As a direct consequence, German engineers began experimenting with fireproof barges and in doing so actually extended the story's life. At Fecamp, in the Seine-Maritime

---

10 Lashmar P & Oliver J, *Britain's Secret Propaganda War*, Sutton Publishing Ltd, 1998, 15.
11 Garnett D, *The Secret History of the PWE* The Political Warfare Executive, 1939-1945, St Ermins Press, 2002, 214.
12 Richards L, *Whispers of War*, Lulu.com, 2012, 7.

department in Normandy, they had filled a fireproof barge with troops and deliberately steered it into a pool of burning water. All died and burnt German bodies washed ashore, further reinforcing the sib. Delmar, ever the opportunist, immediately broadcast to his German listeners:

> "We English, as you know, are notoriously bad
> at languages, and so it will be best, meine Herren
> Engellandfahrer, if you learn a few useful English phrases
> before visiting us. For your first lesson we will take:
> Die Kanaluberfahrt. The Channel crossing, the Channel
> crossing. I will give you a verb that should come in
> useful. Again, please repeat after me,
>
> Ich brenne, I burn.
> Du brennst, you burn.
> Er brennt, he burns.
> Wir brennen, we burn.
> Ihr brennt, you are burning.
>
> And if I may be allowed to suggest a phrase: Der SS
> Sturmfiihrer brennt auch ganz schon, The SS Captain is
> also burning quite nicely, the SS Captain is also burning
> quite nicely!"

For all the innovation of the channel story, many of PWEs sibs were less thoughtful. A great many ideas for sibs were simply outrageous and, had they been deployed, would have been counterproductive. Some of the ideas floated drew criticism – which extended to criticism of the wider PWE – from across the UK government, particularly the Ministry of Information and the War Office. Perhaps the best example was the suggestion that a rumour should be spread that man-eating sharks had been sent from Australia and released into the channel to help resist invading forces![13] However, in November 1944 General Eisenhower wrote to the Combined Chiefs of Staff of the need to 'break the German will to resist'. This chimed with Prime Minster Winston Churchill's 'desire to get at German morale by underground methods'.[14] Or as one historian put it, it was "the search for a cheap method to end the war, one that would make the gruesome battles

---

13 Cumner S, War with the gloves off. The PWE and Lessons Learned from IO, DIS NIAT 3.
14 The National Archives, CAB 120/853, 828 Prime Minister to President 24th November 1944.

on the ground unnecessary".[15] PWE was one of the weapons of choice in the battle for 'hearts and minds'.

PWE was not the only government organisation working these problems. In 1941 The War Cabinet authorised Operation Hard Boiled, a strategic deception operation that was designed to convince Hitler that the allies would soon invade German-occupied Norway. In fact, the operation was more a dry run to evaluate if the UK had the capability to mount large scale deception operations. Enthused by its apparent success, General Wavell lobbied the Prime Minister to establish a bespoke organization that was capable of: "bold and imaginative deception ... by officers with special qualifications". The PM agreed and The London Controlling Section (LCS) was born and Lieutenant Colonel John Bevan made its first Head. Bevan was an excellent choice. He had been involved in tactical deception operations in Norway in 1940 and had a distinguished record from the first war. Unfortunately, beneath him was appointed a committee that included: 'an Indian civil servant, an actor, a soap factory manager and the famous novelist Dennis Wheatley',[16] which many historians have suggested made it significantly less capable.

The LCS had a turbulent war, in many cases struggling to make its mark. However, it was to be its central role in Operation Bodyguard – the strategic plan for allied deception – and in particular its work on Operation Fortitude which was to forever secure its place in the history books. Fortitude was designed to ensure that the Germans would not increase their troop presence in Normandy, which it achieved by promoting the appearance that the Allied forces would attack German positions elsewhere. Equally important was to delay the movement of German reserves to the Normandy beachhead and prevent a potentially disastrous counterattack. The plan aimed to convince the Germans that additional assaults were planned – specifically in Scandinavia and in the Pas de Calais in France. Fortitude necessitated the creation of two fake field armies (based in Edinburgh in Scotland and on the south coast of England) which threatened Norway (Fortitude North) and Pas de Calais (Fortitude South). The operation was intended to divert German attention away from Normandy and, after the invasion on 6th June 1944, to delay reinforcement by convincing the Germans that the Normandy landings were purely a diversionary attack.

---

15 Ambrose, S. E., The Supreme Commander: The War Years of General Dwight D. Eisenhower, Doubleday and Co, 1970, 544.
16 Latimer J, *Deception in War*, John Murray Publishers 2001, 148.

The invasion of Europe and the steps taken to hide its true aim were to be LCS' biggest wartime success.

LCS was not an IO organisation *per se* but inevitably it used information to deceive the Germans and it may well be the genesis of the idea that IO is inherently deceptive and untrue. It is certainly at odds with today's definition of *'accurate and attributable'* but in its defence it was established in the midst of a war of severe existential threat and national survival.

LCS did not end on cessation of hostilities in 1945. Embedded within the Chiefs of Staff organization in the War Office it existed under the name LCS until at least December 1950 and quite possibly later. British Chiefs of Staff Directive 787 issued on 30 November 1950, established three theatre deception offices in the Far East specifically to deal with the threat posed to British interests by the Malayan communist party. However, the threat posed by the Soviet Union was never far from Whitehall's collective mind and in December that year an internal memo[17] noted that: "we have underestimated the scope which exists for misleading the other side about our intentions and exaggerating our strengths"; it later noted that: "we may be able to exaggerate the mobility of our forces in an emergency and thus go some way to compensate for numerical inferiority". Whilst the utility of the functions provided by LCS were not in doubt, its name was. Some felt that the name gave too much of an indication about the nature of its work whilst the use of 'London' in the title implied a strict central control which could hinder its relationship with the local commands in the Far and Middle East. After much thought the LCS was officially re-designated the Directorate of Forward Plans (DFP) in December 1950.[18]

The DFP, which had it not been for political sensitivities would have been called the Directorate of Deceptive Plans, was established under a civilian Director (its first, John Drew, had learnt his trade during Operation CROSSBOW – the code name of the World War II campaign of Anglo-American operations against the German long-range weapons programme), and an Army Deputy (Colonel H N Wild who, as part of LCS, had been responsible for Operation BODYGUARD) and three deception planners, one from each of the armed service. Its primary function was: 'to maintain within the Joint Planning Organisation a deception Planning Team, responsible, under the joint direction of the DFP and of the Directors

---

17 Ministry of Defence, COS (50) 520 12 dated 13 Dec 1950.
18 *British IO since the Second World War.* A paper for DIS by Information Exploitation Analysis Branch dated November 2001.

of Plans for preparing deception plans in support of military strategy in peace and war'.

Although subsequently mandated as an instrument of the Cold War, the DFP slowly found itself drawn into wider emerging areas of IO, often (inaccurately) referred to as Political Warfare, around the British Empire. Palestine, Malaya and Kenya all served to remind Whitehall that future operations would not just be directed against the Soviet Union and in the absence of any other capable organization, DFP took control of PsyOps policy – which paradoxically was ultimately to contribute to its downfall.

Through the 1950s and early 1960s, DFP was busy plying its trade but did so largely in the Middle and Far East and against separatist movements and insurgencies rather than the Soviet army. Indeed, as NATO grew in strength, and greater reliance was placed upon the nuclear deterrent, the requirement for deception planning in the European theatre became less and less relevant. So too the requirement for PsyOps which became a victim of self-fulfilling prophecy; whilst local PsyOps units were maintained in the Middle and the Far East they were small and because of their limited capacity were seen as being ineffective. As was shown in the preamble, 'for the sake of economy [we] run PsyOps on a shoestring basis'. Army Commanders faced with a choice between a PsyOps capability and conventional weaponry chose the latter every time. And why would they not? The PsyOps capability had been allowed to decline. In 1968 the DFP, which looked increasingly moribund, was taken as a cost saving measure and the UK lost not only its deception capability but so too the only champion and coordinator for PsyOps. It was to take the invasion of Kuwait by Iraqi Forces under Saddam Hussein in 1991, and the success of US PsyOps, to rekindle interest in the UK military for an IO capability.

Whilst the MoD may have lost interest, other UK government departments had not but it was not until the 27th January 1978 that a 'new' capability was to be revealed by The Guardian newspaper, which an intriguing story entitled *'Death of a Department that never was'*.[19] Its author, David Leigh, recounted the 30-year history of the Foreign Office's covert propaganda operation. The article announced the demise of the Information Research Department (the IRD), a department it claimed was a worldwide British propaganda network, operating against communism but one which had 'failed to change with the times' and had established

---

19 Leigh D, *Death of the department that never was*, The Guardian Friday 27 January 1978, available at https://www.cambridgeclarion.org/e/fo_deceit_unit_graun_27jan1978.html.

dubious links with certain right-wing journalists. The reason for its demise is more complex than the paper states, what is not in debate is the IRD's part in the Cold War.

The end of the hostilities in Europe saw the eastern nations obscured by Churchill's so-called 'Iron Curtain'. Through the late 1940 and 50s the former allies became less and less trusting of each other's intentions. The Berlin Airlift of 1948 showed the depth of that mistrust but so too the West's intent to sustain and spread ideas of liberal democracy vice communism. In 1949 NATO was established, in 1955 the Warsaw Pact was born and in 1962 the Cuban missile crisis demonstrated to the world just how real the threat of war between the superpowers could be. Perhaps because Cuba so nearly triggered a global Armageddon the superpowers found that fighting their wars through proxy conflicts became a preferred option, principally in Southeast Asia and the Middle East. Quite aside from the threats posed by the Cold War, the UK also found itself embroiled in a series of colonial and post-colonial crises, notably the Malayan Emergency, the Kenyan Mau-Mau uprising, the Suez Crisis, Cyprus and Belize to name just a few. For a country left almost bankrupt after World War II each presented HMG significant challenges. One of its solutions was the creation in 1948, under Prime Minister Clement Atlee, of the IRD.

The Soviet Army had, for the duration of the war, been part of the allied alliance against the Nazis and the war in the east provided breathing space to the US and UK to plan D-Day. The War machine had thus done much to publicize the success of the soviet war effort and to garner support – financial, logistical, and metaphysical – amongst the western British and American populations. 'Suddenly', however, the soviets had become the enemy; former German communists now released from Nazi detention and concentration camps were beginning to take up positions of responsibility in the newly emerging German Democratic Republic (GDR) whilst former, but now de-nazified, Wehrmacht officers were taking up positions in the new Bonn based Federal Republic of Germany. A curious turn of affairs. As Andrew Defty notes: "The IRD could not hope to manipulate a consensus regarding the soviet threat … without the change in public perceptions brought about by events such as the communist subversion of democracy in Czechoslovakia".[20]

By 1973, from its Great George Street office in Westminster, 43 staff, under the supervision of Gordon Reddaway, worked to propagate the

---

20 Defty A, *Britain, America & Anti-Communist Propaganda 1945-53*, Routledge 2004. 5.

west's view on communism. The IRD researched, created, and distributed propaganda material to British and foreign recipients in positions of influence. In his analysis of the IRD, Simon Collier wrote: "Much of this propaganda was banal, factual information; sometimes facts were twisted – 'spun' – and occasionally facts were stretched or massaged, but there is little evidence that IRD promulgated straight falsehoods".[21]

Perhaps surprisingly the biggest single section of IRD, at least in May 1973, was not that concerned with the Soviet Union and Eastern Europe (four staff) but International Organisations (five staff). The Middle East and North Africa merited only two, whilst the Far East and Southeast Asia only one desk officer respectively.[22] The work was almost exclusively un-attributable, but it was truthful and accurate. One of the main outlets was the quarterly magazine 'Communist Affairs'. The contents page of Issue number 256 from October 1974 perfectly indicates the range of IRDs interest: Section 1 Forthcoming events (World Student Conference on Chile, National Congress of the Young Communist League); Section 2: Communist Party Issues; Section 3: The Trade Union Movement; Section 4: The Peace Movement; Section 5: Youth Movements; Section 6: Women's Movements; Section 7: Friendship Societies. By any stretch of imagination this had now moved on from the original intent of defeating war time adversities and is undoubtedly why the IRD – and the organisation that was to follow it – latterly attracted such negative attention.

In October 2007, the British government announced the establishment of the Research, Information and Communication Unit (RICU), a counter-terrorism strategic communications unit run from the Home Office. It exists to ensure that the Government has a positive impact in its counter-terrorist communications, and to both counter the impact of terrorist communications directly and assist others in doing so. According to the Home Office, RICU has three key deliverables: first, advising CONTEST[23] partners on counter-terrorism related communications; second, exposing the weaknesses of violent extremists' ideologies and brand; and last, supporting credible alternatives to violent extremism. RICU has a staff of around 30 people and reports to three different ministers of state – the

---

21 Collier S M W, *Countering Communist and Nasserite Propaganda: the Foreign Office Information Research Department in the Middle East and Africa, 1954-1963*, University of Hertfordshire, 2014. Available at https://uhra.herts.ac.uk/bitstream/handle/2299/14327/04085529%20Collier%20Simon-%20Final%20PhD%20submission.pdf.
22 IRD Staff List dated 14 May 1973 and held in MoD archives.
23 The UK's national counter-terrorism strategy is known as 'CONTEST' and is based on a policy that can be summed up in four words: Prevent, Pursue, Protect and Prepare.

Home Secretary, the Minister for Government and Local Communities, and the Foreign Secretary. Notably not the Secretary of State for Defence and the MoD.

Between May 2007 and June 2007, Dr Jamie Macintosh – who became my 'boss' in the Advanced Research Group shortly after – was personal adviser to the Home Secretary, John Reid, for transformation and national security. He was in part responsible for RICU's creation, which he states was a response to a perceived absence of capacity in central government. He recalled the Home Secretary's thinking: "The Prime Minister had a Lieutenant for every other portfolio of government, except the first duty of government, national security. And that meant that the Prime Minister himself stretched across every portfolio of government and so the first duty was not getting the attention it should have. Blair agreed with that analysis."[24] Macintosh recalled that: "The only other time in UK history when these matters had been properly addressed and integrated was the Cabinet War Rooms of Churchill and in particular the role of PWE ... the Home Secretary understood that we needed a PWE if we were serious about strategy making, and engaging in the battle of ideas, seamlessly from operational theatres abroad to the home front."[25]

Back in the MoD, in March 2000, the Chiefs of Staff Committee met to discuss the future policy of MoD IO. The Chiefs had been particularly interested in the application of IO during the operations in Kosovo the preceding year, the chiefs concluded that: "the potential of Info Ops has been recognised".[26] The Chiefs went on to note that there were: "potential difficulties in the coordination of targeting and info ops... [intend] to create a new Directorate of Targeting and IO (DTIO)". In its Terms of Reference, the new Directorate was to provide a focal point in the MoD for Targeting and IO in peace, crisis and conflict and to provide the necessary integration between the Commitments and Intelligence Areas. Alongside Chairmanship of a whole host of new committees and targeting boards the new 1* Director[27] was to supervise all MoD Targeting and IO effort and develop operational requirements for IO and targeting. In selecting the new 1* Director the MoD decided they would need an: "officer from a broad operational background.

---

24 Private discussions between Dr Jamie Macintosh and author, ARAG, June 2009.
25 Ibid.
26 *A Policy Framework for IO* CDS 121/00 dated 31 March 2000.
27 In most Armed Forces a Star (*) system is used to designate senior officers. A Brigadier (or RN/RAF equivalent) is a 1*, a Major General (or equivalent) is a 2*, a Lieutenant General (or equivalent) a 3* and a full General a 4*. This system is also used in certain Civil Service positions with Senior Civil Service (SCS) grades assigned comparable stars.

It is not a rotational post but should be filled by competitive selection across all three services – experience of Joint Ops is essential and experience of targeting and intelligence matters desirable". Chief of Defence Intelligence (CDi) and Deputy Chief of Defence Staff (Commitments) (DCDS(C)) were both DTIO's boss and the new organisation was located in the historic Old War Office building across the road from MoD Main Building (and now a smart London hotel!). In December 2008, as a response to cross government policy which elevated the title of Director from the 1* to 2* level, DTIO had to change its name to TIO. In yet another departmental organisation in 2012, after involvement in the operations against the Libyan regime in 2011, TIO became part of the Operations Directorate and relocated from the Old War Office to the 4th floor of MoD Main Building on Whitehall and renamed Military Strategic Effects (MSE).

But for all this strategic and policy level development, an almost constant historical feature of UK efforts has actually been the urge of commanders and management boards to delete their operational structures from the inventory of defence – and then hastily attempt to regenerate when 'the balloon has gone up' and the capability and expertise largely gone. In the midst of the Suez crisis in 1956, for example, the Commander in Chief of Middle East Land Forces, and commander of Operation MUSKETEER,[28] General Sir Charles Keightly, wrote that: "PsyOps is an essential part of military planning... I recommend that there should be a really 1st Class PsyOps Warfare Cadre always in being"; During the crisis in Cyprus in 1958 the command recalled that "PsyOps could not be mounted quickly enough or sustained effectively on an ad hoc basis"; in Anguilla in 1969 the Senior Naval Officer for the West Indies reflected that "one lesson stood out clearly from this operation. We had no means of disseminating information and countering adverse local propaganda"; in British Honduras in 1970 it was reported back to London that "we need a capability to encourage popular support for the security forces, to persuade the uncommitted and encourage anti-Guatemalan sentiment". As the MoD's files reveal, the list of plaintive cries for specialist information support is long and repetitive.[29]

Yet these pleas, from in-theatre commanders, were invariably met with indifference in Whitehall. Take, for example, a December 1965 letter from Major General J Harrington, HQ Far East Command, Singapore, to

---

28 Operation Musketeer was the Anglo-French plan for the invasion and capture of the Suez Canal.
29 These selected quotes have been extracted from unreferenced documentation held in the Ministry of Defence archives.

Permanent Under Secretary MoD recommending the establishment of a small PsyOps capability in Borneo. His request was supported by Deputy Director of Forward Plans who penned: "we strongly support this request … the Chiefs of Staff recently instructed us to intensify the war of nerves in the Far East. HQ Far East has for the sake of economy run PsyOps on a shoestring basis". On the 1st February 1966, Brigadier GS from Military Operations (1), MoD, replied: "There is no manpower cover for such a proposal without compensating reductions. This cannot be justified". In March 1967, Commander in Chief (CinC) Far East apparently gave up the fight and wrote: "In present financial climate there was no option but to recommend that the plans for the formation of a PsyOps Unit should be dropped". It is interesting to note that in 1967 there were 406,600 service men and women[30] (For comparison at the 1st February 2021, the UK Armed Forces were 150,000)[31] and that defence spending, as a percentage of GDP was 6.77 per cent.[32] Today we are well used to hearing of budget and workforce constraints, that any increase in capability must be met with a commensurate reduction elsewhere. One might be forgiven from asking if the armed forces really were genuinely constrained in 1967 or if the request was declined simply because the capability was so poorly understood?

Because of the length of British military operations in Northern Ireland, the scale of literature related to the campaign is almost overwhelming and most remains classified; that which specifically concerns IO especially so. In part this is due to the fact that the entire information campaign was devolved to Belfast but also because much of the information campaign was associated with Special Forces and Intelligence operations. However some sense of IO activity is now publicly available in '*An Analysis of Military Operations in Northern Ireland*' Army Code 71842 dated 2006.[33] Signed off by the then Chief of the General Staff, General Sir Mike Jackson, the report states:

> For several reasons IO were probably the most
> disappointing aspect of the campaign. IO were generally

---

30 *Army cuts: how have UK armed forces personnel numbers changed over time?*, The Guardian, 1 Sep 2011, available at http://www.guardian.co.uk/news/datablog/2011/sep/01/military-service-personnel-total.
31 *UK defence personnel statistics*, House of Commons Reference Library, available at https://commonslibrary.parliament.uk/research-briefings/cbp-7930/.
32 Ibid.
33 *Operation Banner. An Analysis of Military Operations in Northern Ireland*, Army Code 71842, available at https://www.vilaweb.cat/media/attach/vwedts/docs/op_banner_analysis_released.pdf.

poorly conducted; they were ill-coordinated with other government bodies; they were reactive; and often missed significant opportunities. The absence of a government information line was often exploited by the terrorist, sometimes with operational or strategic consequences. Constant criticism in the republican media, notably the An Phoblacht newspaper, was not seriously challenged by Government, NIO or Army IO. Part of the reason for the ineffectiveness lay in the lack of a single unitary authority for the campaign, and the lack of a joint forum to agree IO priorities, messages and means of dissemination. Differing viewpoints on the need for positive IO in Stormont (and the NIO) the Police and HQ Northern Ireland militated against effective IO.

This is not to say that there were not successes. The now famous 'Cats in the Cradle' video[34] was described by The Belfast Telegraph as the: "most powerful advert ever to hit our TV screens".[35] The video told the story of a father and son caught up in the sectarian violence of the Seventies and Eighties and when first aired on Ulster TV in 1993 it created a significant reaction amongst the Northern Ireland population. David Lyle, who wrote it, told the Telegraph that: "What we were trying to get across was that terrorism wrecks families… it spoke deeply and powerfully to people … we ended it 'Don't suffer it, change it'. And, of course, the change eventually happened."[36] Largely overlooked by many commentators is the fact that IO during Operation Banner and any attempts to conduct PsyOps would be targeted against sections of the UK population and, as such, meet significant doctrinal, legal, and ethical constraints.

Following the 1991 Gulf War, when US Psychological Operations had been seen to work very well in persuading Iraqi soldiers to surrender the UK formed what it referred to as a shadow PsyOps capability. It settled on the name 15 (UK) PsyOps Group (Shadow), the number 15 was chosen as PsyOps activities during World War II were conducted by Radio Amplifier

---

34 *Cats in The Cradle Northern Ireland add*, available at https://www.youtube.com/watch?v=3i_p1mupPio.
35 Rutherford A, *Eerily prophetic Troubles ad that shocked us in 1993 gets 500,000 views in one day*, The Belfast Telegraph, 30 June 2016, available at https://www.belfasttelegraph.co.uk/news/northern-ireland/eerily-prophetic-troubles-ad-that-shocked-us-in-1993-gets-500000-views-in-one-day-34407900.html.
36 Ibid.

Units numbered 10-14, and the group adopted the stags head (with antennae like antlers) formation sign in recognition of the work carried out by the Indian Field Broadcast Units that supported the Chindit campaign against the Japanese in Southeast Asia. It was not until 1998 however that the unit lost the 'shadow' (as in not formerly established, rather than any nefarious meaning – often misinterpreted by some) and was properly stood up; for the next 16 years it supported UK operations globally, its battle honours covering Kosovo, Sierra Leone, Macedonia, Afghanistan, Iraq, and Libya. Unhappily it held the distinction of being home to the first female British soldier to die on operations in Afghanistan, Corporal Sarah Bryant, while serving alongside the SAS Reserve, in 2008

15 PsyOps – which I had the privilege to command for nearly three years – was deployed in every major UK military operation but by far its longest was Operation Herrick in Afghanistan. 15 PsyOps deployed to Helmand in 2005 and had been the centre piece of a slow change in military tactics and thinking; bombs and bullets may win short term skirmishes and battles they made poor ambassadors for long term success and peace, particularly in a country where conflict and violence was the norm to an entire generation and where expectations of peace and normalcy are almost non-existent. The PsyOps team set out to change that perspective. In the light of the ignominious withdrawal of western forces in August 2021, it is easy to assume that everything was a failure but that was simply not the case. 15 PsyOps activities were significant in reducing support for Taliban and opening people's minds to other possibilities for their nation. Broadly their activities panned the traditional audience understanding of the rich and very varied human tapestry that characterised Helmand, trying to build up a detailed corpus of social and cultural knowledge of the province's people in order to promote trusting relationships through non-kinetic activities. Concurrently the group assisted education and training programmes to British and other International Stabilisation and Assistance Forces (ISAF) units to facilitate such deeper understanding of the perceptions and motivations of the civilian population that conventional force was not always regarded as the default action.

In 2014, just as in previous decades, the MoD decided to amalgamate its 'soft effects' units or 'Influence Orphans' (those that sought to achieve outcomes without recourse to violence) into one central organisation; 15 PsyOps, together with the Media Operations Group (Volunteers) an entirely Reservist specialist unit, and the Military Stabilisation Support Group (MSSG), merged initially into the Security Assistance group (SAG) before that in turn was renamed the 77th Brigade and homed in Hermitage in Berkshire. Around a similar time, another much smaller and very much tactically focussed unit was established by the Royal Marines – 30 Commando Information Exploitation Group designed to assist 3 Commando Brigade with IO.

77th Brigade had a rocky start. For a supposedly specialist information unit, it managed to court public controversy immediately when it adopted the badge of the former World War II Special Forces Group, The Chindits, without informing the Chindits Association in advance! 77th Brigade slowly grew in size into a large and increasingly capable unit; one which like 15 PsyOps Group before it has proven irresistible to both the mainstream media and social media conspiracists alike. In January 2023, for example, an army whistle-blower told The Daily Mail newspaper that 77X had been engaged in monitoring British public social media activity during the COVID antivax crisis of the previous years.[37] To be fair to the Army, the story was missing some key information, not least that the bulk of work was being undertaken by a civilian company working under a contract with the Department of Culture, Media and Sport and, critically, whatever work 77X was doing was not for the MoD but for the Cabinet Office under long standing MACA arrangements.[38] This is important; the allegation that the British Army was 'spying' on the British public was simply scaremongering. What 77X were doing was monitoring on behalf of the government, in the midst of a national emergency, mis and dis information appearing online. However the story played badly. The Whistleblower told the newspaper: "It was about domestic perception, not national security. Frankly, the work I was doing should never have happened. This domestic monitoring of citizens seemed not to be driven by a desire to address the public's concerns, but to identify levers for compliance with controversial Government policies."[39] The Unit will always attract conspiracy theorists;

---

37 *Army spied on lockdown critics: Sceptics, including our own Peter Hitchens, long suspected they were under surveillance. Now we've obtained official records that prove they were right all along*, The Daily Mail, 29 Jan 2023, available at https://www.dailymail.co.uk/news/article-11687675/Army-spied-lockdown-critics-Sceptics-including-Peter-Hitchens-suspected-watched.html.

38 MACA is Military Aid to the Civil Authorities. To quote from UK government documents, Military aid is usually called upon to provide capabilities which the relevant civil authorities do not have, or cannot generate in sufficient quantity in response to an emergency. At the national level, military personnel work with colleagues across relevant Government departments to plan for and act in the event of an emergency. See UK Government Fact Sheet 14, Military Aid to Civil Authorites available at https://assets.publishing.service.gov.uk/media/5a795d6b40f0b642860d779f/Factsheet14-Military-Aid-Civil-Authorities.pdf

39 Ibid.

whilst there was some small grain of truth in the Daily Mail's article there is plenty of commentary about 77X that is entirely fictional. For example, it has been accused of rigging the Scottish independence vote (which actually occurred long before the unit was formed!)[40] and it is often featured in rants on Russian TV channel RT.

But why, when there is such obvious and growing need for units capable of operating in the information domain world is IO and PsyOps even considered as a capability that can be cut when resources are tight? In part this might be due to the discreet nature of the deployment of information campaigns; the absence of 'boots on the ground' does not automatically mean the absence of IO but because of the inevitable political sensitivity of such operations knowledge of their deployment is invariably placed on close hold. However, this argument might equally be applied to intelligence operations and neither UK domestic intelligence agencies nor the Army Intelligence Corps have faced such existential threat, indeed both have expanded in the post-9/11 years.

Firstly, and unlike conventional military operations, there is no immediate and 'satisfying' effect. A properly structured campaign may take months to mature and whilst measures of performance (number of leaflets produced, hours of broadcast of a radio station) make it comparatively easy to measure the campaign's activity – the elusive and far more useful Measure of Effect (MOE) is significantly less easy to determine. Correlation and causality bedevil such campaigns and when money is tight – as the files amply reveal – these small points matter. However, there is also a more pedestrian and far less edifying explanation. The fact is that senior officers' careers are not built on IO but on tanks, aircraft, and ships; their previous exposure to what are often called the 'dark arts' is often minimal *ergo* their understanding and interest; in over 25 years of IO service I can count on one hand the number of very senior officers who truly recognised the value of IO and who were prepared to afford it priority over kinetics.

Secondly – cost. As former principal advisor to Secretary General of NATO, Chis Donnelly CMG, wrote in his private Defence Academy paper, *'The realities of Defence Economics'*: "For 300 years, during peacetime, the English / UK Defence Budget has been remarkably consistent at between 2-5 per cent of GDP. At 2-3 per cent GDP – without the running costs of

---

40 *Leading lawyer offers to represent mad Nationalist if he proves 'Brits' were behind his followers dropping on Twitter*, The Scottish Daily Express, 9 Nov 2022, available at https://www.scottishdailyexpress.co.uk/news/politics/leading-lawyer-offers-represent-mad-28453120.

current operations, we cannot sustain the capability to conduct the full spectrum of military operations that we have in the past. To do that we would need 4-5 per cent GDP. Even to maintain our current reduced capabilities and associated minimal structures, Defence needs more money than it is getting. Either we need a serious increase in the Defence Budget or we need to introduce drastic changes in the way we do things".

The irony is that IO projects can be surprisingly cost effective – particularly when used in advance of conflict as a deterrence – yet the absence of knowledge about them means they are often seen as 'nice to have' as opposed to being essential. As will be seen later, IO campaigns work; the evidence base stretches from contemporary times back as far as you wish to explore. Even if the leadership of the MoD was not persuaded by this, the House of Commons Defence select committee was. In its Third Report of Session on Lessons of Iraq it declared that: "We are persuaded that IO are an activity which can be expected to become of increasing importance in future operations. There were a number of successes which provide evidence of the potential effectiveness of IO. We recommend that the Government should consider significantly enhancing our capabilities in this area".[41]

Perhaps because of Iraq and Afghanistan, finally, a significant change to UK military strategy and thinking occurred in 2020 when the MoD announced its Multi-Domain Integration change programme, a massive cross defence programme to break down single service and departmental barriers and generate an improved ability to operate, at pace, in an era of persistent competition and in sub war fighting threshold scenarios. Or, in other words, to generate the ability to confront Russian, Chinese and non-state actors in Information and Cyber Operations. What that will look like remains to be seen. Perhaps, bizarrely, the failure of the UK mission in Afghanistan which I believe was largely a result of failing to understand population groups – specifically the Afghan National Army – might actually positively reinforce the need for sustained, funded and enduring UK IO capability underpinned by a robust 'understand' process (which I address in a later chapter). And, certainly, the 2022 war in Ukraine has opened the eyes of even the most sceptical leaders to the power of information.

---

41 *Information Operations*. Select Committee on Defence Third Report, available at http://www.publications.parliament.uk/pa/cm200304/cmselect/cmdfence/635/63506.htm. paragraph 465.

3 The USA – Bigger and Better?

In 2003, when British troops went to war in Iraq, stories emerged about the poor state of preparedness of British servicemen and women's equipment. Leading British tabloid, The Daily Mail, told its readers that US troops had begun referring to their British colleagues as 'The Borrowers' because of their "habitual scrounging for basic equipment, or to replace their own inferior kit with American issue".[1] I was certainly pleased to visit the on-base 'Exchange' (known as 'the PX' and essentially a department type store for troops) to trade my 1980s manufactured and now rapidly disintegrating desert combat boots for modern US ones! Through the years UK equipment has greatly improved but I will never forget that first sense of wonderment at working alongside US forces and seeing how well resourced and how well equipped they were.

For that, they have the US defence budget to thank, which in FY2022/2023 was nearly $773 billion a year[2] (for comparison the UK budget for 2023 is £48billion or $57 billion) – that is bigger than the next nine global defence budgets in total. As a result, the size and scale of the US military is almost impossible to describe in any meaningful or detailed way; it is simply enormous. Comprising the Navy, Army, Air Force, Marine Corps, Coast Guard and Space Force, over 1.4 million men and women wear US military uniform,[3] although that does not mean the US military, in personnel, is the largest – that 'honour' falls to China which is reckoned to have around two million men and women under arms.

If not in training, then almost all US Military personnel are attached to one of the 11 global combatant commands (COCOM); such as Central Command (invariably abbreviated to CENTCOM and set up to specifically

---

1 Pendlebury R, *Our poor bloody infantry*, The Daily Mail, 2 Oct 2006, available at https://www.dailymail.co.uk/news/article-407954/Our-poor-bloody-infantry.html.
2 *The Department of Defense Releases the President's Fiscal Year 2023 Defense Budget*, US DOD Press Release, 28 March 2022, available at https://www.defense.gov/News/Releases/Release/Article/2980014/the-department-of-defense-releases-the-presidents-fiscal-year-2023-defense-budg/.
3 *Armed forces of the United States – Statistics & Facts*, Statista.com, available at https://www.statista.com/topics/2171/armed-forces-of-the-united-states/.

cover the middle east), Special Operations Command (SOCOM), Europe Command (EUCOM), and Indo-Pacific command (INDOPACOM). The 11 commands are overseen by the Unified Command Plan, or UCP. The UCP defines each command's mission, aligning its outputs to US national policy objectives. Each is commanded by a four-star flag officer, who answers directly to the US Secretary of Defense and each COCOM commander can come from any branch of military service.

In keeping with its huge size and resources, IO is also vast in US doctrine. In my introductory chapters I bemoaned the 'tyranny of terminology'; and chose to use the generic term IO as a 'catch all' throughout this book. In this chapter, alone, I need to momentarily walk back from that because the US regards the term IO merely as an integrating or planning function – and indeed has a cadre of military personnel who do nothing but – whereas the actual delivery of information related activity is undertaken by specially trained personnel who, in contrast to the UK, spend the entirety of their military careers working in this highly specialised area. And in the US, PsyOps is referred to as PSYOP!

The US defines IO as: 'the integrated employment, during military operations, to influence, disrupt, corrupt, or usurp the decision making of adversaries and potential adversaries while protecting our own'. However as the former Commander of US Strategic Command, General Kevin Chilton, commented, "You ask ten different people what IO is, and you'll get ten completely different answers".[4] And like the UK, the US has been adept at describing the problem. The latest doctrinal term is 'information advantage' which holds the distinction of being the seventh iteration in Army IO terminology in less than four decades. First there was 'command, control, and communications countermeasures' in 1981, then 'command and control warfare', 'information warfare', 'IO', 'inform and influence activities', until today's 'information advantage'. As Professor Sally White, an Army cyberspace operations officer and former assistant professor of international affairs at West Point wrote: "the Army has spent the greater part of the past 40 years trying to determine what information is and what it means for warfighting. The service has cycled and recycled through numerous attempts at doctrinal codification, and yet it seems to be in much

---

4 McFate M, Military Anthropology: Soldiers, Scholars and Subjects at the Margins of Empire, C Hurst & Co Publishers Ltd, 2018.

the same place in 2023 as it was in the early 1990s."[5] All of which sounds remarkably like the UK's experience.

Whilst the US may be wrestling with doctrine and definitions it is at least funding IO activities. In 2023 the US Department of Defense (DOD) announced that US Army IO activities would receive $238.8 million in funding, which is nearly $40 million more than they had requested, with most of that money being spent in INDOPACOM[6] – an indication perhaps of who the US regards likely as the most important audiences in the future. To put that into perspective, the most I have ever seen spent on UK military IO was £50 million, and that was a budget allocation to be spent over five years.

Perhaps the most visible part of the DOD's IO output is the work undertaken by its Psychological Operations units. Overseeing that effort are the 4th and the 8th Psychological Operations Groups (POG), both based at Fort Bragg in North Carolina. Beneath the 4th POG sits the 3rd, 6th, 7th, and 8th Psychological Operations Battalions and beneath the 8th POG sits the 1st, 5th, and 9th Battalions – all serving different US Army areas of operations. In the Army reserve the 2nd POG has eight PSYOP battalions.[7] In total, maybe 200-230 in the training school, around 3000 reservists and approximately 2500 active duty; in total nearly 6000 PSYOP personnel compared with the UK, which at the height of operations in Afghanistan, had around 120 personnel in 15 (UK) PsyOps Group. With the possible (and largely unknown) exception of China, this is the largest professional concentration of military PSYOP personnel anywhere in the world and it means that any deploying US Army unit, globally, can have access to PSYOP support if needed.

The term PSYOPs has itself become contested in recent years. PSYOP's name changed to Military Information Support Operations (MISO) around 2012 largely because of concern over the sinister syntax.[8] As Wired Magazine

---

5 White S, *The Organizational Determinants of Military Doctrine: A History of Army IO*, Texas National Security Review, Vol 6, Iss 1 Winter 2022/2023, available at https://tnsr.org/2023/01/the-organizational-determinants-of-military-doctrine-a-history-of-army-information-operations/.
6 Williams L, *More Money For Info Ops, Army Recruiting, Cyber In Omnibus*, Defense One, 22 December 2022, available at https://www.defenseone.com/defense-systems/2022/12/more-money-info-ops-army-recruiting-cyber-omnibus/381278.
7 *US Army active Duty PSYOP units*, Psychological Operations Veteran's Association, available at https://www.usapova.org/us-army-psyop-units.
8 Ambinder M, *Original Document: Making PSYOPS Less Sinister*, The Atlantic, 30 June 2010, available at https://www.theatlantic.com/politics/archive/2010/06/original-document-making-psyops-less-sinister/58947/.

exclaimed, "Psychological Operations sounds awesomely creepy",[9] which is perhaps a more relatable explanation for the change than that offered by the DOD which told the world that: "The name change has been an emotional topic and has been bandied about for years… Lack of emphasis of influence operations by senior combat CDR and the bifurcation of PSYOP forces between SOF and Reserve chains of command continues to hobble efforts to optimize and standardize PSYOP training and operations"![10] Curiously though the name reverted to PSYOP in 2014, back to MISO in 2015 and today is still collectively and colloquially know across the service as … PSYOP.

US Psychological Operations were allegedly first used by American troops during the Revolutionary War, when leaflets calling for enemy soldiers to desert the British military were distributed. The US Library of Congress reports that the leaflets had some success and some British servicemen deserted, subsequently settling permanently in America. They are also recorded to have been used in the Civil War when allegedly large balloons with the pictures of Union Generals were floated across Confederate positions.

Most discussion about Psychological Operations is focussed on its active element – the influencing of foreign audiences and adversaries. However, during World War I, the US government became concerned that the US population might become unduly influenced by other nations – particularly Germany. In part this was recognition at just how rural the nation still was and news, such as it was over 100 years ago, could take days or weeks to arrive and invariably when it did it would be transmitted via a local filter – be it word of mouth or local newspapers. This 'last three feet' was what the then President, Woodrow Wilson, sought to address when he established the Committee on Public Information (CPI). Echoing some of the problems of mis- and dis information being gripped by democracies today, the US President saw the need for some kind of psychological defence against malign information directed at the population.

Between April 1917 and June 1919, the CPI acted as an independent agency of the US government and used a variety of methods to reach the

---

9 Shachtman N, *Military Mulls New Name for Psychological Operations: MISO*, Wired, 25 June 2010, available at https://www.wired.com/2010/06/military-mulls-new-name-of-psychological-ops-miso/.
10 *MISO: Is it soup yet?* Phil Taylor's Papers, Leeds University School of Communications, available at https://universityofleeds.github.io/philtaylorpapers/vp010117.html.

US public, most famously the so called 'Four Minute Men',[11] volunteers coached to deliver a concise government message to public gatherings within four minutes. By the end of World War I it is thought that some 75,000 Four Minute Men speakers had been recruited; delivering over 755,000 speeches to audiences exceeding, in total, more than 314 million Americans.[12] The CPI, which latterly became colloquially known as the Creel Committee after George Creel was appointed its head, was actively working not just to counter foreign influence but to specifically influence US public opinion to support the US' entry into World War I.[13] Critically that included encouraging US citizens to boost food production, to save much needed war materials and, critically, to encourage enlistment in the armed services. The CPI was responsible for one of the most iconic posters of all time – Uncle Sam pointing towards the viewer and captioned *'I want You for The US Army'* – created by James Montgomery Flagg. It is an image that endures, today, over 100 years later.

The coming of World War II saw PSYOP coming under the control of the Office of Strategic Services (OSS). The OSS was led by Willian Donovan who today is recognised as the founder of the CIA. Working closely with the UK's Political Warfare Executive (PWE), the OSS worked across most of the theatres of conflict, although notably not in the Pacific Theatre where General Douglas MacArthur, the Commanding General, had no time for PSYOP and who was hostile to the OSS. This contrasted with the Supreme Commander in Europe, General Dwight D. Eisenhower, who had been impressed by the work of a small PSYOP team during the allied landings in French North Africa. He initiated the establishment of the Psychological Warfare division into his HQ, the Supreme Headquarters Allied Expeditionary Force (SHAEF), charged with responsibility for waging psychological warfare across continental Europe and placed under the command of Brigadier General Robert McClure.[14]

Throughout the campaign PSYOP were regularly launched. Operation Huguenot was a plan to undermine the efficiency of the German Airforce

---

11 Leo W, *The History of US Psychological Operations: World War One*, SOFREP, 19 March 2020, available at https://sofrep.com/news/the-history-of-u-s-psychological-operations-precursor-in-the-great-war-part-2/.
12 Cornebise A, *War as Advertised. The Four Minute Men and America's Crusade, 1917-1918*, Philadelphia : American Philosophical Society, 1984.
13 Maxwell C, *George Creel and the Committee on Public Information 1917*-1918, Tenor of Our Times Volume 4 Spring 2015.
14 *The Psychological Warfare Division SHAEF, An Account for Operations in the Western European Campaign 1944-1045*, reprinted by Lightning source UK.

by providing Luftwaffe commanders with intercepted 'evidence' indicating high rates of desertion in German aircraft to the allied side; Operation Nest Egg was a (never used) plan to accelerate the surrender of the Channel Islands and centred on using a captured German General to contact the Garrison Commander on the islands to induce him to surrender; Operation Braddock II was an operation directed at foreign workers inside Germany and created concern in the minds of the German authorities that they were an internal threat to security inside the fatherland[15]; Operation Sauerkraut[16] was initiated by OSS within hours of the failed assassination attempt on Adolf Hitler in 1944. Designed to exploit growing disillusionment within the German Army, the operation attempted to encourage more German troops to follow Count von Stauffenberg's example and attack the Nazi leadership. Some German prisoners in Italy were deemed to be particularly valuable and were released back to Germany to sow seeds of discontent. Carrying thousands of leaflets, the soldiers distributed them wherever they could; one announced that Field Marshall Albert Kesselring had resigned out of fear that the Nazis were losing the war,[17] apparently forcing Kesselring to issue a formal address refuting the rumour. Because of the perceived success of these operations, US PSYOP were expanded throughout the European theatre of operations.

Following World War II and post the North Korean invasion of the South in 1950, President Truman created the Psychological Strategy Board (PSB), to handle the US' national PSYOP policy,[18] and in the same year, General McClure of SHAEF fame, was named Chief of the Psychological Warfare Division. McClure and his staff established the PSYWAR Center at Fort Bragg in 1952[19] and 11 officers were assigned to the Army's newly reorganized Psychological Warfare Division.[20] Major Albert Brauer – one of those 11 – recalled that the first and greatest problem in conducting psychological warfare in Korea was: "to know the mind and character of the

---

15 Ibid, 41.
16 *Operation Sauerkraut*, Codenames.info, available at https://codenames.info/operation/sauerkraut/.
17 Fratus M, *Killer Vampires, Demon Dolls, And Sauerkraut: A Brief History Of American Psyops*, Coffeeordie.com, available at https://coffeeordie.com/craziest-american-psyops/.
18 Friedman H, The American Psyop Organization During The Korean War, Psywarrior.com, available at https://www.psywarrior.com/KoreaPSYOPHist.html.
19 *Homepage 4th POG*, US Army website, available at https://www.soc.mil/4thPOG/4thPOGhome.html.
20 Brauer A, *Psychological Warfare Korea 1951*, Georgetown University, available at https://library.ndsu.edu/ndsuarchives/sites/default/files/KoreanWarPSYCHOLOGICAL-WARFARE-KOREA19513.pdf

target audience....against an oriental target it is especially difficult because of the wide differences between the cultures of the East and the West".[21] The Korean war saw the mass use of propaganda leaflets, particularly ones designed to influence the Communist enemy to defect or simply leave the North Korean Army. However the complexity of the audience and low literacy rates amongst north Korean soldiers meant the leaflets' success was mixed.[22]

It was another war in Asia that was to turbo-charge the US' development of Psychological Operations – Vietnam – which has been described as "the most sustained, intensive use of psychological operations in American history".[23] Vietnam would see numerous PSYOP programs running. By its end, a four-station radio network had been created, with the ability to access 95 per cent of the population 24 hours a day; a four-transmitter TV network had been established with six hours of programmes per day and as many as 50 billion leaflets had been distributed in an area the size of California. Like in Korea new structures were created, one being the 6th Psychological Operations Battalion, which was initiated specifically for operations in Vietnam in November 1965. At its peak, the total number of military personnel committed to the PSYOP campaign was 1200 US military and just over 750 Vietnamese personnel, operating on an annual budget of over $12 million (the equivalent of $80 million in 2023).[24]

Yet for all its investment and personnel, the US PSYOP campaign in Vietnam whilst large, was not regarded as being particularly effective. In lessons resonant of today, PSYOP were seen as ancillary to conventional military force and thus always the second choice for commanders on the ground. Secondly, "the US sought to substitute its own PSYOP programme for that of its Vietnamese ally",[25] a mistake that ISAF forces tried very hard not to make in Afghanistan. Finally, the US lacked a national level organisation for the command and control of PSYOPs and as a result its efforts across the country were fragmented and lacked coordination (something that NATO today tried to avoid with the deployment of a Combined Joint PsyOps

---

21 Brauer A, *Op.Cit.*
22 Ross F. Collins & Andrew D. Pritchard (2016) *Pictures From the Sky: Propaganda Leaflet Psyop During the Korean War,* Visual Communication Quarterly, 23:4, 210-222.
23 Roberts M, *Let the dogs bark: the psychological war in Vietnam, 1960-1968,* available at https://digital.library.unt.edu/ark:/67531/metadc849646/m2/1/high_res_d/ROBERTS-DISSERTATION-2016.pdf.
24 Lord C, *Political Warfare and Psychological Operations. Rethinking the US Approach.* National Defence University Press, 1989, 123-124.
25 Ibid, 125.

Task Force to enduing campaigns). Despite these shortcomings, Vietnam would provide the backdrop for a PSYOPs campaign that has now entered the annuls of IO history – the use of ghosts to haunt Viet Cong soldiers. Named "Operation Wandering Soul", the campaign was designed to erode Vietnamese moral through the broadcasting into the jungle of screaming and shrieking human voices. The soldiers who heard the voices, uncertain of their source, would, it was hoped, be paralyzed with fear.[26] Did it work? As with all IO the issue of correlation and causality bedevils accurate assessment. There were a number of Viet Cong defections during the time it was in effect but was that as a direct result of the 'Ghosts' or attributable to other factors?

Today the US regular Army's principal Psychological Operations capability is vested in 4th POG Group and between the Vietnam and Afghanistan wars, 4th POG has participated in operations in Vietnam, the Dominican Republic, Panama, Iraq, Somalia, Haiti, Liberia, Rwanda, Bosnia, and the Balkans.[27] Between December 1990 and February 1991 more than 29 million PSYOP leaflets were disseminated by 4th POG operators in Iraq; 66 tactical PSYOP loud speaker teams attached themselves to advancing US units and were successful in persuading thousands of Iraqi soldiers to surrender; the 'Voice of the Gulf' radio station was broadcast continuously from 19 January until the cessation of hostilities, targeted specifically at Iraqi soldiers.[28] Many of those radio broadcasts were made from a specially modified Hercules aircraft – the so called 'Commando Solo', which could broadcast on digital and analogue radio and, if needed, television broadcasts. Flown by the 193rd Special Operations Wing of the Pennsylvania Air National Guard unit, Commando Solo only finally left service in September 2022.[29]

The Gulf War was good for both UK and US PSYOP. The UK was impressed at the way US PSYOP were able to encourage the mass surrender of Iraqi forces. Major Peter Whiteneck, a US Marine, wrote his staff college paper on PSYOP in Iraq and concluded that "PSYOPS [in Iraq]

---

26 Fitzgerald C, *Operation Wandering Soul: The US Military's Use of Psychological Warfare in Vietnam*, War History Online, August 2022, available at https://www.warhistoryonline.com/vietnam-war/operation-wandering-soul.html?safari=1.
27 *4th Psychological Operations Group website*, DOD, available at https://www.soc.mil/4thPOG/4thPOGhome.html.
28 *Leaflets of the Persian Gulf War*, 4th POG, US DOD Published Brochure.
29 *Farewell to the EC-130J Commando Solo III, the plane of the USAF for psychological operations*, Defensa Aviacion, 19 Sept 2022, available at https://www.outono.net/elentir/2022/09/19/farewell-to-the-ec-130j-commando-solo-iii-the-plane-of-the-usaf-for-psychological-operations/.

was not conducted according to doctrine, it was nevertheless effective in all mediums, significantly contributing to mass capitulation of Iraqi forces and an overwhelming victory for the Coalition".[30] Operation Desert Storm – as the US referred to the conflict – also demonstrated the importance of global mass media, which for the first time was covered, live, 24 hours a day, by international media and resulted in the US Army undertaking the "greatest hands-on application of media relations ever."[31] It was the success of the US information campaign that encouraged the UK to ultimately establish 15 (UK) PsyOps Group as described in the previous chapter.

Throughout operations in Afghanistan, PSYOPs were used extensively by US troops. The US was one of only three troop contributing nations (TCN) able to provide the staffing for the Combined Joint PsyOps Task Force Headquarters (CJPOTF) in Kabul, the overall in-theatre coordinating body; the others were Germany and Romania but all three would augment their staffs with PSYOP officers seconded from other NATO nations. Given the length of the campaign, over 20 years, there is now a huge literature base that examines almost every aspect of the operation. One study that received quite wide attention was the RAND corporation think-tank's assessment of US PSYOP. Given the amount of money invested, the huge number of people deployed and the multinational nature of the operation, RAND is far from glowing in its assessment, noting that "The most-notable shortcoming has been in countering the Taliban's propaganda campaign against U.S. and coalition activity, which has focused on civilian casualties and has found a broad national and international audience".[32] Unlike the obvious success of the first Gulf War, RAND believed that the Afghan PSYOP campaign (of which I was part and in which I commanded the UK PsyOps Group in Helmand) struggled to make its mark and observed that the biggest successes had been in the areas of "face-to-face communication and meetings with key communicators, such as local councils of elders, local leaders, and members of the Afghan media".[33] This is perhaps unfair. The campaign was so wide ranging – from the pursuit of Bin Laden and AQ terrorists in Tora Bora in December 2001, through to encouraging Afghans

---

30 Whitenack P, *An Analysis Of Gulf War Psyops And Their Applicability To Future Operations*, School of Media and Communications, available at https://universityofleeds.github.io/philtaylorpapers/vp01a6a8-2.html.
31 White S, *Op.Cit.*
32 Munzo A, *U.S. Military IO in Afghanistan Effectiveness of Psychological Operations 2001-2010*, The Rand Corporation, 2012, available at https://www.rand.org/pubs/monographs/MG1060.html.
33 Ibid.

to vote in nation-wide elections – that blanket assessments are unhelpful. From my vantage point in Helmand there were many operational successes of PSYOPs, but it is clear, given the ignominious pull out in 2021, ultimately, we did not prevail and PSYOPs must, like every other element of the coalition operation, take its share of the blame.

So far this chapter has almost exclusively focussed on the US Army but it is of note that the US Marine Corps has comparatively recently joined the US IO 'team' with the establishment of the Marine Corps IO center (MCIOC). Established in 2009 in Quantico, Virginia, the MICOC became fully operationally capable in February 2011. MICOC exists to support US Marine Expeditionary Forces and units supporting them with "subject matter experts, teams and detachments, and psychological operators".[34] Much like its US Army sister organisation, MICOC provides regionally focussed support: Team 1 looks after the Middle East, Team 2 the Pacific and South America, and Team 3 focuses on Europe and Africa. MICOC chooses to focus its work on six capability areas: working within the electromagnetic spectrum, cyber, space, targeted influence operations, deception operations and inform operations.

In recognition of the accelerating importance of IO, in 2022 Congress directed that the US DOD set up a new senior leadership role, mirroring the success of a similar appointment for the management of defense Cyber capabilities two years previously. Major General Matt Easley became the first Deputy Principal IO advisor to the Secretary of State for Defense. At the London Phoenix Challenge conference in April 2023[35] he explained his four key missions: To create a lexicon for IO that would be understood across the US DOD; to understand exactly what IO capabilities the Department of Defense possessed (that in itself an admission of the chaotic management of IO to date); to devise a strategy for US IO going forward; and to establish a joint IO force provider and trainer. Or as he told briefing attendees "I'm trying to organise the Pentagon to deal with IO".[36]

Finally, it would be remiss not to briefly mention other US Information Operation type organisations, outside of the DOD. The most obvious is the US State Department's Global Engagement Centre (GEC) which leads

---

34 Chawk F, *Marine Corps IO Centre*, Marine Corps Gazette 2020, available at https://mca-marines.org/wp-content/uploads/0420-Marine-Corps-Information-Operations-Center.pdf.
35 Phoenix Challenge is a bi-annual conference for Western IO practitioners; April 2023 was held in London. See https://defensescoop.com/2023/06/16/inside-phoenix-challenge-the-conference-series-seeking-to-shape-and-bolster-information-operations/.
36 Private briefing to Five Eyes IO community at Exercise Phoenix Challenge, London February 2023.

the US Federal Government's efforts to understand and counter foreign state and non-state propaganda (arguably not dissimilar to Woodrow Wilson's CPI of 1919). The GEC lists six specific missions on its website: Analytics and Research, International Partnerships, Programs and Campaigns, Exposure, (of foreign influence operations) and Technology Assessment and Engagement.

Less obvious are the Information Operation's undertaken by the US' Central Intelligence Agency (CIA). Perhaps the most visible component of that was the OSC – the Open Source Centre. Established in 2005, OSC was tasked to collect and analyse open-source information of intelligence value across all media. The OSC changed its name to the Open Source Enterprise (OSE) in 2015, a CIA spokesperson briefing that the OSE is "dedicated to collecting, analysing, and disseminating publicly available information of intelligence value".[37]

As for more 'active' IO, like all intelligence organisations it is all but impossible to accurately verify current and recent activities. The agencies own, public, website is however interesting in that it allows access to archive documents that appear to link the Agency with PSYOP[38] and disinformation, at least historically. There are, for example, open-source references to operations in Nicaragua and Libya.[39] There are interviews on YouTube with alleged past CIA employees, one talks about his role in spreading disinformation about North Vietnamese effort to develop airfields in order to persuade the US Congress that it should continue aid to the South Vietnamese.[40] Perhaps what is most interesting about that interview is that the former agent concludes by saying not only is he now opposed to disinformation but that "it served no useful purpose". A google search reveals countless websites offering theories, some more conspiratorial than others, on how the CIA has used PSYOP to overthrow different governments. You can even buy the, supposed, CIA Manual of Psychological Operations on Amazon![41]

---

37 Aftergood S, *Open Source Center (OSC) Becomes Open Source Enterprise (OSE)*, Federation of American Scientists, 28 Oct 2015, available at https://fas.org/blogs/secrecy/2015/10/osc-ose/.
38 *The Nature of Psychological Warfare*, 1953, available at https://www.cia.gov/readingroom/docs/CIA-RDP81-01043R002400220001-6.pdf.
39 *Disinformation. An examination of six years of incredible lying*, Los Angeles Weekly 1987, available at https://www.cia.gov/readingroom/docs/CIA-RDP90-00965R000807550012-7.pdf.
40 Formally available at https://www.youtube.com/watch?v=aemyhNJUAz.
41 Nagy A, *CIA: Manual for Psychological Operations In Guerrilla Warfare*, CreateSpace, 2011, available at https://www.amazon.co.uk/Cia-PSYCHOLOGICAL-OPERATIONS-GUERRILLA-WARFARE/dp/1466238356.

We do know that in 2016, just before leaving office, President Obama issued a Presidential Finding that authorised the Agency to engage in non-lethal covert action against Russia – the result of the US Intelligence Community finding that Russia had tried to interfere with the 2016 election.[42] In the December 2022 Podcast 'Doomsday Watch' former British diplomat Arthur Snell speaks with ex-US Special Forces soldier and now US investigative journalist, Jack Murphy. Murphy speaks about covert action and how the agency has different organisations within to manage them, including disinformation. How much of this, if any at all, is true and how much falls into the realm of conspiracy theories is mere speculation but it does not seem unreasonable to presume that covert IO may be being undertaken, particularly in the Ukraine / Russia conflict.

What can be said, definitively, is that across the array of US government departments, information is regarded as a key tool and at not just the tactical level, where US Army PSYOP tend to work, but also at the strategic level. And as such information is better understood, better funded and better resourced than in the UK and Europe.

---

[42] See former UK Diplomat Arthur Snell's Doomsdaywatch.co.uk interview with Jack Murphy 11 January 2023, *'Who's burning Russia'*, available at https://www.patreon.com/doomsdaywatch.

4 Iraq and Afghanistan

For most of the armed forces – not just in the UK but across the NATO alliance – experience of IO will most likely have come from service in either Iraq (mainly British and US forces) or Afghanistan (which at its peak had personnel from NATO's 28-member countries and 22 partners).[1] The conflict in Iraq – which the UK named Operation TELIC and the US named Operation IRAQI FREEDOM – lasted from 2003 to 2011 (although its after effects continue to reverberate today) and the Afghanistan Conflict, named Operation HERRICK by the UK and ENDURING FREEDOM by the US, from 2001 until 2021. Both self-evidently tested the US led coalition to its fullest – not just in its application of conventional kinetic firepower but so too its use of IO. Whilst the extent of the failure in Afghanistan became globally apparent in August 2021, with the appalling scenes of Afghans attempting to flee the country and falling from the wheels of military aircraft as they took off from Kabul airport, for some inside the armed forces it almost came as a shock, so deep was the belief that the coalition would succeed. In their assessment of the withdrawal from Kabul the Whitehouse issued a briefing paper in which they declared that 'in early 2021 ... the intelligence and military consensus was that the Afghan National Security Forces (ANSF) would be able to effectively fight to defend their country and their capital, Kabul. The ANSF had significant advantages. Compared to the Taliban, they had vastly superior numbers and equipment: 300,000 troops compared to 80,000 Taliban fighters, an air force, and two decades of training and support.'[2]

Yet as early as 2013 many of us had misgivings and indeed that year I and my immediate superior in the UK Ministry of Defence suggested to the then Strategy Director in Kabul that we should undertake some research into what behavioural influences might lead to the ANSF – the collective noun for the Afghan National Army (ANA) and the Afghan

---
1 *NATO and Afghanistan*, NATO website, 2022, available at https://www.nato.int/cps/en/natohq/topics_8189.htm.
2 US Withdrawal from Afghanistan. A Whitehouse briefing paper available at https://www.whitehouse.gov/wp-content/uploads/2023/04/US-Withdrawal-from-Afghanistan.pdf.

National Police (ANP) – from fracturing and falling apart. The Director was not pleased and dismissed the idea out of hand. He seemed unimpressed at what he regarded as a defeatist attitude. But the evidence that things were going wrong was growing. Right across the country incidents between ANSF, between ANSF and NATO and perhaps most disturbingly, clear fraternisation between ANSF and known Taliban, were being reported again and again. We had also seen evidence that the numbers of ANSF fighters was being significantly inflated, whilst the numbers leaving was being played down. The Director's hubris was by no means unique. In their assessment of the totality of the mission the Dutch military noted a "collective wishful thinking emerged in which staff within the NATO organisation and in participating countries stuck to the same positive narrative even though the evidence did not support this". It would have taken a very brave soul, knowing how unpopular they would become, to suggest in Kabul or in NATO HQ that the (by then) 12 years of hard work to build the Afghan forces could fall apart at any moment. But fall apart it did, dramatically, and the first signs of that were absolutely evident as early as 2013.

It is my view, although not that of the UK MoD today, that the absence of understanding of the human factor – why ANA and ANP did what they did – was a significant contributor to the fall of Kabul in 2021. My views are not unique. Take for example the well-respected Foreign Affairs Journal which reported in August 2021: "the most striking American misjudgement is our ongoing overestimation of the capabilities of the Afghan National Defense and Security Forces. Even without tactical American military support, the ANDSF should have been in a position to defend major cities and critical military installations... in 2017 and again in 2019, there were reports that tens of thousands of 'ghost' soldiers (soldiers who existed on paperwork, particularly the paperwork that accrued money, but nowhere else) were being removed from the rolls – suggesting that there were never close to 330,000 troops available to fight the Taliban, let alone 352,000".[3] The Defense Department's December 2020 report to Congress noted that only "approximately 298,000 ANSF personnel were eligible for pay,"[4] hinting at the recurring problem with 'ghost' soldiers and desertions.

---
3 Lamb C, *Chronicle of a Defeat Foretold Why America Failed in Afghanistan*, Foreign Affairs, July/August 2021 available at https://www.foreignaffairs.com/reviews/review-essay/2021-06-22/chronicle-defeat-foretold.
4 *Why The Afghan Security Forces Collapsed*, SIGAR 23-16-IP EVALUATION REPORT, February 2023, available at https://www.sigar.mil/pdf/evaluations/SIGAR-23-16-IP.pdf.

The issue of understanding – or rather its absence – is perhaps best illustrated by a campaign anecdote. Male village elders had approached a western provincial reconstruction team (PRT), of which there were 27 across Afghanistan, operated by at least 13 different countries,[5] and asked that the PRT sink them a new well in the village. The PRTs were staffed by a mix of military Civil Military Co-operation personnel (CIMIC), international development staff from various nations, and contractors and NGOs. Their purpose was to assist local governance efforts and bring positive change to local areas in support of the wider international military effort against the Taliban. The sinking of a well was therefore a perfectly valid and seemingly self-evidently humanitarian action; what could go wrong? The anecdote, that despite its wide circulation has proven difficult to find a definitive source for, is that shortly after sinking the well the male village elders returned to the PRT and asked for assistance as the well had been poisoned. The PRT returned and cleaned the well yet twice more the male elders return and explained that it had been poisoned. The assumption was that the Taliban were responsible but no evidence of Taliban activity in the area was found. It is only after further investigation that it was decided it was not the Taliban but the village women folk who had poisoned the well. The women, it transpires, were not consulted in the decision to sink the well and were not unhappy to escape the confines of the village men folk and have time together for themselves on their, up until now, daily trek with donkeys and plastic water containers to fetch water.

The moral of the story – and hence its prevalence – is that you must understand the complexity of the environment, and its many different stakeholders, if you hope to make any positive difference. It is unfortunately the case that today's armed forces are likely to have only the most cursory understanding of the environments in which they are called to work despite the wall of evidence from past conflicts that this is utterly seminal to mission success. This absence of understanding was, arguably, to define the failure of the West's operations in both Afghanistan and Iraq.

Former Commander of Coalition troops in Afghanistan, General Stanley McChrystal, told the US Council on Foreign Relations that: "The US and its NATO allies are only 'a little better than half way' to achieving their military goals, partly due to a frighteningly simplistic understanding of the

---

5 *Provincial Reconstruction Teams (PRTS)*, Institute for the Study of War, available at http://www.understandingwar.org/provincial-reconstruction-teams-prts.

country."⁶ Matt Cavanagh, a former advisor to the British Government, told the Journal of the Royal United Services Institute that: "No one inside the British [government] system knew much about the insurgency, the opium trade or the local politics and tribal dynamics – or just as importantly about how these different elements fitted into each other. Planners and policy makers did not know much about the human or even physical geography of Helmand"⁷ whilst the late Michael Hastings wrote in his book *The Operators* that: "In one meeting [Andrew] Exum drills down on the briefers. Who controls the water? Who are the local power brokers? Tell me how they are related to the insurgency? The Intel Officers shrug. The questions 'scare the hell out of them'".⁸

Yet a non-military reader might well be surprised at these admissions; even if they don't understand the details, it is widely known that the armed forces have huge intelligence collection organizations. The number of *'ints'* is substantial. OSINT (Open Source Intelligence); HUMINT (Human Intelligence); GEOINT (geo-spatial intelligence); SIGINT (Signals Intelligence); IMINT (Imagery Intelligence); MARINT (Maritime Intelligence) – and the list goes on. But for all of these capabilities the wars in Iraq and Afghanistan more than amply demonstrated that one critical source of intelligence is missing – what I have come to refer to as POPINT, or population intelligence. The simple fact is that most military intelligence is aimed at the adversary and the adversaries' equipment not at understanding the wider population. Toward the end of operations in Afghanistan, the amount of data being collected on a daily basis was astonishing. In just one intelligence sub-organization over a terabyte of data was collected each week⁹ and yet, as is now painfully obvious to everyone, the entire mission to Afghanistan failed and the Taliban once again control the country.

Like the urinal anecdote that appears in Chapter 9, the well-sinking anecdote is a useful educational example for IO because it shows that, under certain circumstances, it is possible to predict behaviours. Is behavioural prediction something that the military would benefit from? The answer must be 'yes'. The military's preference is generally to deter violent

---

6 Speech to Council on Foreign Relations July 2011.
7 *Ministerial Decision-Making in the Run-Up to the Helmand Deployment*, RUSI Journal May / June 2012, available at https://rusi.org/explore-our-research/publications/rusi-journal/ministerial-decision-making-run-helmand-deployment.
8 Hastings M, *The Operators: The Wild and Terrifying Inside Story of America's War in Afghanistan*, Little, Brown & Company, 2011.
9 Interview Tatham / UK Intelligence Officer and former Officer Commanding specialist Intelligence collection in Helmand, August 2021.

behaviour but it exists, ultimately, to legally deliver extreme violence against adversaries. To be able to predict what circumstances might lead to violent behaviour, and when and where it might occur, and to do so with a degree of confidence if not absolute certainty, is an incredibly powerful tool and one that is absolutely predicated on understanding audiences.

Iraq

In mid 2002 I was told that I had to report to the secure underground headquarters in Northwood – home to the Permanent Joint Head Quarters – to attend a brief and be signed on to something that at the time was referred to as OPLAN 1003 Victor. I had no idea until I received that brief that planning for the invasion of Iraq was so advanced. From then on I was tasked with developing the Royal Navy's Strategic Communication plan for the invasion and, after being seconded to the UK's Maritime Command in Bahrain, subsequently ended up as one of the public spokesmen for that invasion. In 2006 I wrote of my experiences of dealing with the Arab media in '*Losing Arab Hearts and Minds*', where the absence of nuanced understanding of audiences again made our task of communicating very difficult. Regardless, during the entire period of planning for the invasion I recall never once hearing any mention of someone called Muqtada al-Sadr and his (Shia) Mahdi Army. Instead the focus was on the Sunni tribes of Iraq for the presumption was that the Shia, so oppressed under Saddam Hussein, would naturally welcome the 'liberating' forces. It was a shock to everyone, therefore, when after the fall of Baghdad, it was the Mahdi Army and their long-exiled leader al-Sadr, now returned from Iran, who caused the UK so much difficulty in southern Iraq. Once again, we had failed to understand the population.

As early as 2004, US military recognised the need for IO in Iraq and some of the challenges they would face – not least in the understanding of the different component audiences. In an article for the Military Review, *IO in Iraq*, Major Norman Emery wrote that "Success requires comprehending the intricacies of the Iraqi psyche—the tribal loyalties, the stubborn sense of national pride, the painfully learned distrust of America's promises … the United States must convince Iraqis that the temporary US military presence in Iraq is necessary to rebuild the country for the benefit of the Iraqi people".[10] Yet quickly that aspiration to inform, persuade and positively

---

10 Emery N, *IO in Iraq*, The Military Review May-June 2004, available at https://universityofleeds.github.io/philtaylorpapers/pmt/exhibits/1619/ioIraq.pdf.

influence was lost as a full-blown insurgency began. One of the principal weapons of the insurgent was the Improvised Explosive Device (IED).

The US' attempts to reduce the effect of IEDs in Iraq is an excellent example of how the need to thoroughly understand audiences before embarking on IO is so necessary to the military. Between 2004 and 2007 the rate of IED attacks in Iraq increased dramatically. In January 2004 there were around 500 IED incidents resulting in around 300 deaths and/or injuries. By November 2004 IED incidents were approaching 1500 a month and deaths or injuries around 700. By May 2007 IED were peaking at nearly 4000 per month with a corresponding number of deaths or injuries at around 1000 per month.[11] Behind these statistics were stories of human suffering – dead and horribly maimed soldiers and, because of their indiscriminate nature, Iraqi civilians too caught up in the blasts.

Unsurprisingly the statistics soon became an issue of presidential interest and the US stood up the Joint IED Defeat Organization (JIEEDO), tasked to come up with quick and successful solutions to dealing with the IED threat. Amongst millions of dollars of investment were mobile surveillance systems for force protection, route clearance blowers, specialist bomb suits, robotic systems, specialist armour for vehicles, X-ray machines, specialist data bases for collection and analysis, personal protective equipment and an allocation for a non-attributable multi-media influence effort owned by the US Special Operations Command, as well as more general IO reach back support.[12] Very little exists in the public domain about either but within the IO community another anecdotal case study has existed for many years and it involves a commercial public relations agency that was successful in gaining a major US strategic communication contract in Iraq to support JIEEDO.

Anecdotally an assumption was made that IEDs were being laid because Iraqis were unable to relate to US service personnel as humans; rather, behind their Kevlar body armour, helmets and weapons were actual human beings with families and children and not simply automated killing machines. And so, a series of videos, leaflets and billboards were made that portrayed the humanity of US service personnel, of them with their families and children, many set against well-known US backdrops such as the State of Liberty. The idea was that Iraqi would 'like' what they saw,

---

11 *Monthly Frequency of Improvised Explosive Device (IED) Incidents and Combat Deaths and Injuries, 2004–2009* taken from Mil Med. 2017 Mar; 182(3): e1697–e1703, available at https://www.ncbi.nlm.nih.gov/pmc/articles/PMC5459305/.
12 *JIEDDO Annual report FY 2009*, available at https://www.hsdl.org/?view&did=682217.

and this might dissuade them from laying IEDs – an attempt at perception management.

The campaign certainly had an effect; within a few weeks the number of IED explosions went up. This caused a rethink and proper behavioural analysis was undertaken to determine the motivation for laying IEDs. And a new, and crucial, piece of information was uncovered. It transpired that in the majority of cases the people laying the IEDs were not the same people as those making the IEDs. For those ideologically driven IED 'layers' there was very little that could be done to dissuade them. But the bomb makers were different. For many the motivation to assemble IEDs was money. And they wanted to make as much as they could for a simple and understandable reason. Iraqi society had largely fallen apart. Sectarian death squads roamed; electricity was frequently off (and so too air conditioning in the sweltering heat of the Iraq summer), water supplies intermittent, and rubbish and sewage was collecting in the streets. With high unemployment and long-term prospects poor it seems entirely understandable that people would wish to leave – which needed money. And a quick source of revenue was manufacturing IEDs. So far so obvious but what the analysis revealed next is what really surprised people, the place that a majority of Iraqis wanted to get to – the USA. The USA was a land of freedom and wealth and safety and opportunity and for a huge number of people it was the reason they made IEDs. They saw no contradiction between their actions and their aspirations and the images of US families set against backdrops of the Statue of Liberty and waving wheat fields in Kansas simply strengthened those aspirations.

The 'solution' (in quotes since in complicated and complex human eco-systems there is often no single solution) was to replace the imagery of happy Americans with horrific images of dead and mutilated Iraqi children, the other victims of IED attack. Slowly the rate of IEDs went down. Was this entirely a result of the IO? No. It would be disingenuous to say it was as with so many other JIEDDO initiatives going on correlation and causation would have been difficult to differentiate but certainly a more audience centric approach that understood the motivations for behaviours was better than the original 'guess' that was made and which might even have cost lives.

Unfortunately, naivety was to characterise many of the coalition's IO efforts in both Iraq and Afghanistan and both became rich pickings for civilian communication companies. In 2016, former Bell Pottinger cameraman, Martin Wells, gave an interview to The Times newspaper in which he spoke about his role in filming and editing videos in Iraq.[13] His work, he told the paper, included creating short news videos, which he made to look like proper Arabic news clips, as well as fake insurgent videos, which would be burnt onto CDs and dropped into areas that were raided. The CDs had a code embedded in them that provided the location of where they had been subsequently played. PR Week magazine reported that "Bell Pottinger made about £15 million a year in fees from the work"[14] whilst agency founder Lord Bell told The Sunday Times. "I mean if you look at the situation now, it wouldn't appear to have worked but at the time, who knows, if it saved one life it [was] a good thing to do." Bell Pottinger conducted its work under the IO Task Force (IOTF), and the scale of their budgets was extraordinary. Journalist at US Today, Tom Van den Brook, has tracked the US' expenditure. In a 2013 newspaper piece[15] he wrote that: "Propaganda spending …: $9 million in 2005 to $580 million in 2009. Its recent budgets: $488 million in 2010; $355 million in 2011; $202 million in 2012."

In 2007 the US DOD, recognising that it had only the most limited understanding of the human tapestry of tribes and groups across Iraq, introduced the Human Terrain System (HTS), deployed anthropologists and psychologists.[16] Attached to military units across Iraq (and latterly Afghanistan) the deployed Social Scientists attempted to understand the insurgency better. At its cessation in 2014, HTS had cost almost $750 million making it the "largest investment in a single social science project in U.S. government history".[17] It was a lot of money, but against the backdrop of the wider costs of the military operation it should have been seen as good value because it offered commanders insight into an area of operations where they were effectively blinded – the population – in which the insurgency was raging. Professor Montgomery McFate who started the programme told me that: "HTS was shut down when the US pulled out of Iraq, there

---

13 Harrington J, *Bell Pottinger in the spotlight for creating propaganda videos for US military in Iraq*, PR Week, available at https://www.prweek.com/article/1410858/bell-pottinger-spotlight-creating-propaganda-videos-us-military-iraq.
14 Ibid.
15 Vanden Brook T, *Propaganda or IO: Words matter*, USAToday, 8 Nov 2013, available at https://eu.usatoday.com/story/nation/2013/08/11/pentagon-propaganda-military-information-support-operations/2635509/.
16 The UK had a smaller but not dissimilar programme called Cultural Advisors (CulAds). At their heart the HTS and CULADs were seeking to develop an understanding of the population eco-system.
17 Sims C, *Academics in Foxholes. The life and death of the Human Terrain System*, Foreign Affairs, 4 February 2016, available at https://www.foreignaffairs.com/articles/afghanistan/2016-02-04/academics-foxholes.

was no need for the program any more".¹⁸ But she also hinted at a wider problem. HTS had been bombarded by criticism from US academics and commentators and HTS personnel were becoming worried that they might not actually find jobs back in academia and research in the future. McFate told me that "Most academic anthropologists don't understand how the military works and what exactly 'counterinsurgency' means. The objective is to reduce the level of violence and increase the functioning of government. One important means to do that is to increase the military's level of understanding of the local social-political-economic system. But academic anthropologists always seemed to think it was just a sneaky way to kill more civilians". Or as Zenia Helbig, a former HTS operator, told the media: "They [academic community] just didn't want to hear anything that didn't jive with their conspiracy theories".¹⁹ As I will show later, the actions of some elements of the academic community on both sides of the Atlantic have placed military IO programmes under stress.

It was not just the US led forces who were busy with IO. The combined Iraqi and Al-Qaeda opposition was also busy in the information environment. Al-Qaeda had always understood the need to mobilise Muslim populations – the 9/11 attacks were not just a physical strike on the US but also a massive IO designed to galvanise, principally, the global Muslim population – the ummah – and demonstrate that the military behemoth that was the US was itself quite vulnerable. Analysis of jihadist communiqués between 2001 and 2005 shows 92 per cent of their output focussed on Muslim audiences.²⁰ Abu Ubeid al-Qurashi, a leading AQ strategist, wrote: "They did not aspire to gain Western sympathy; rather, they sought to expose the American lie and deceit to the peoples of the world – and first and foremost to the Islamic peoples".²¹ But Iraq was to see a change in AQ's output – a switch to targeting western audiences, adapting their output to meet Western tastes, to sow discontent and fear and apply pressure through the lens of communication onto western governments.

An interesting example of this was 'Juba the sniper', allegedly of the Islamic Army of Iraq – who may or may not have existed in reality but who

---

18 Email Tatham / McFate 28 January 2020.
19 Shachtman N, *Academics Turn On "Human Terrain" Whistleblower*, Wired, 3 Dec 2007, available at https://www.wired.com/2007/12/the-fight-betwe/.
20 Torres M, Jordan J and Horsburgh N, *Analysis and Evolution of the Global Jihadist Movement Propaganda*, vol. 18 no.3, Fall, 2006, 399-421.
21 *Al-Qa'ida Activist, Abu 'Ubeid Al Qurashi: Comparing Munich (Olympics) Attack 1972 to September 11 MEMRI*, 12 March 2002, available at https://www.memri.org/reports/al-qaida-activist-abu-ubeid-al-qurashi-comparing-munich-olympics-attack-1972-september-11.

in mythology certainly did. The Washington Post wrote a feature about his exploits in which one US soldier, Travis Burress, himself a sniper based in Camp Rustamiyah near Baghdad, told the media: "He's good… Every time we dismount, I'm sure everyone has got him in the back of their minds. He's a serious threat to us."[22] Juba became a 'hero' of the resistance, his exploits relayed via the internet. A specialist Juba website[23] (now since gone) was set up honouring his achievements containing English commentary and claims of responsibility for the deaths of hundreds of Coalition soldiers. After nearly three years of online activity the legend of 'Juba' appeared to have run its course. Some attributed this to his death, while others say that no one sniper ever actually existed. Regardless, the ability of the story to demoralise coalition troops is significant and reminiscent of the stories of Vasily Zaitsev, the famed soviet sniper of Stalingrad in 1941.[24]

Another significant example of the adversaries IO was an operation called 'Lee's Life for Lies'. This operation involved fabricating a history of US Marine Lee Kendall, whose USB flash drive was found by insurgents.[25] The information became the basis of a series of fake letters and stories that portrayed the apparently desperate and demoralised state of the US soldiers in Iraq and, critically, alleged a series of war crimes. Kendall, it was alleged, was an anti-war activist and 'proved' that many US soldiers serving in Iraq were against the war. A series of downloads were circulated, many of which are still available today on web archive pages.[26]

Between 2005 and 2007, Al-Qaeda's Ayman Zawahiri quadrupled video output,[27] almost all of it on the back of the ongoing conflict in Iraq. Whilst the internet was the easiest medium for distribution, it was the videos that began appearing on the Arabic TV channel Al-Jazeera, with its huge global audiences, that caused the most concern to coalition officials. On 22nd November 2005 the UK's Daily Mirror newspaper reported that "President Bush planned to bomb Arab TV station Al-Jazeera in friendly Qatar, a Top Secret No 10 memo reveals" but it went on to say that the US President was "talked out of it at a White House summit by Tony Blair,

---

22 Taylor A, *Iraqi Sniper: The legendary insurgent who claimed to have killed scores of American troops*, The Washington Post, 22 January 2015, available at https://www.washingtonpost.com/news/worldviews/wp/2015/01/22/iraqi-sniper-the-legendary-insurgent-who-claimed-to-have-killed-scores-of-american-soldiers/.
23 http://www.baghdadsniper.net/.
24 Vasily Zaytsev – The legendary sniper of the Stalingrad battle https://stalingradfront.com/articles/articles-about-stalingrad-battle/zaytsev/.
25 *Lee's Life For Lies*, available at http://www.lee-flash.blogspot.com.
26 *Lee'sLifeForLies*,availableathttps://www.neowin.net/forum/topic/530290-lees-life-for-lies/.
27 Hoffman B, '*Scarier than Bin Laden*', Washington Post, Sunday Outlook section, 9 September 2007.

who said it would provoke a worldwide backlash".²⁸ As one of the coalition public spokesmen at the time this seems highly improbable; Al-Jazeera's HQ was just a few miles from the CENTCOM HQ in Doha, Qatar – an allied nation – and whilst there was considerable lobbying going on behind the scenes to get the station to moderate its coverage, a bombing raid seems very far-fetched. Nevertheless, it served to fuel the hysteria.

Afghanistan

During the Soviet occupation of Afghanistan in the 1980s, two Afghan Islamist insurgent organisations, Gulbuddin Hekmatyar's Hezb-e-Islami and Ahmad Shah Massoud's Jamiat-e-Islami, had both used media campaigns in their military operations against the Soviets.²⁹ With no internet they instead produced inexpensive magazines, local radio broadcasts, newsletters, and occasional video and audio, to promote their cause in Afghanistan and, in particular, in Pakistan. However, these were amateur in design and production and the targeting of their audience sporadic and ill-defined. The Taliban's IO against the US and coalition forces in Afghanistan, post 9/11, were slow to start but in time were to prove more sophisticated although sporadic and, initially, uncoordinated.

The Taliban were not known for their press freedom. Images of human beings were considered apostate and world public opinion largely irrelevant. However, by mid-2002 the Taliban had taken the first steps in the construction of a widespread traditional propaganda campaign that would include "the distribution of dictums, leaflets, cassettes and books that call for jihad and explain the punishment for those who cooperate with or work for the crusaders".³⁰ But, in late 2006 reports began circulating that the Taliban had sent representatives to Iraq to learn how Al-Qaeda's video production arm – now called Al-Sahab (the Cloud) – worked. The following year videos started appearing on the Internet and, of perhaps most significance, in April that year the Taliban allowed a journalist to 'embed' with them – perhaps unsurprisingly it was Al-Jazeera's Pakistan correspondent.³¹

---

28 *U.S.: Al-Jazeera bomb story 'outlandish,* CNN, 23 November 2005, available at https://edition.cnn.com/2005/WORLD/europe/11/22/us.al.jazeera/.
29 *Terrorism Focus*, vol. IV issue 15, 22 May 2007. See also A. Borovik, *The Hidden War: A Russian Journalist's Account of the Soviet War in Afghanistan* (Grove Press, 1990).
30 *Jamestown Terrorism Focus*, vol. 4 issue 10, 17April 2007.
31 *Taliban Media Production Centre*, YouTube, available at http://www.youtube.com/watch?v=E2IjuNYAF4Y

A five-part series, reporting on the experience, was subsequently aired by Al-Jazeera. One episode was entitled *'The People's Movement'* and gave the first indication of a concerted Taliban 'hearts and minds' campaign. In that piece an (alleged) female Afghan doctor declared her support for the Taliban, her blue burqa conspicuously absent, whilst tribal elders spoke with approval of the peace and security that the Taliban had brought.[32]

In June 2007, a video of a Taliban suicide graduation ceremony reached the international media.[33] Self-evidently designed not for a Muslim audience but for a Western viewer, one young man stands up and states: "let me tell you why I will be making a suicide bomb in Britain". Others talk of taking attacks to Ottawa, Canada, and to Germany. Indeed, the Taliban's leader, Mullah Dadullah says: "Listen, all you Westerners and Americans. You came from thousands of kilometres away to fight us. Now we will get back to you in your countries and attack you." The graduation ceremony, which has the appearance of a college graduation, has the 'students' organised in six national 'brigades' (British, American, Canadian, German, French and Afghan) who take turns in pledging future action. A Canadian official told the Canadian Television News channel that they took the Taliban threat 'very seriously',[34] stating that Canadian intelligence had known for some months that the Taliban leadership had directed its commanders to 'take the fight out of the country, to take it to us'. Celebrated Pakistani journalist Hamid Mir confirmed the report as 'absolutely true' – suicide bombers, it would appear, were heading to Canada, America, and Europe.[35]

Whilst elements of the large Pakistani diaspora in the UK might potentially be a proving ground for such operations, the claims that the Taliban and their acolytes might have the capability and resources to cross the Atlantic does seem unlikely. And yet, whether they could or not was largely an academic point for, as with the female doctor in the earlier video, they are fine pieces of directed IO, designed to intimidate and instil fear in Western audiences.

Also in June 2007, the Taliban commander, Mansor Dadullah, provided a long and detailed interview with Al-Jazeera and in July the Taliban announced to the world, again via the conduits of Al-Jazeera and the web, that they have re-branded themselves as 'neo-Taliban'. Key

---

32 *editeddoctorfco*, YouTube, available at https://www.youtube.com/watch?v=YuW4DukwnHk.
33 *PhDresearcher*, You Tube, available at https://www.youtube.com/watch?v=H8RERSoYzpg.
34 Available at http://www.ctv.ca/servlet/ArticleNews/story/CTVNews/20070618/taliban_bombers_070618/20070618.
35 Available at http://www.canadafreepress.com/2007/cover062107.html.

in the channel's three-part documentary was the filming of the Taliban's media centre; sophisticated video editing equipment, much of it in the English language, being used to churn out the Taliban's message.[36] Although not the most sophisticated video-editing equipment, the news feature showed Windows-based software (along with the ubiquitous PowerPoint presentations), in English, being used to create videos and CDs immortalising the Taliban's fight. The piece caused one colleague from the security service's Joint Terrorism and Analysis Centre (JTAC) to speculate that perhaps young radicalised British Muslims were now choosing to fight their personal jihads not with AK-47 assault weapons but with computers and video-editing equipment.

In October, the English-language outlet of Al-Jazeera was sent 14 videos by the Taliban. Among the footage were attacks on Afghan police vehicles. One tape showed a person trying to escape before being shot. The cars are then set alight. Another tape shows Taliban fighters proudly displaying what they discovered at an empty US military outpost. They appear particularly intrigued by the night-vision goggles.[37] As the Al-Jazeera correspondent notes: "The desire to gain the psychological advantage has meant that armed groups here and across the Middle East have now embraced propaganda in a big way. On all sides of the Afghan conflict there is an awareness that while the battles are important, the message may help win the war". Throughout 2008 the momentum was maintained and in autumn of that year the Taliban provided Al-Jazeera International (the English language channel) some 14 video tapes of their operations against coalition troops. In September, the Taliban courted the renowned French magazine *'Paris Match'*,[38] posing for its photographers partially dressed in the uniforms of dead French soldiers, killed only days beforehand. The French Defence Minister Herve Morin asked: "Should we really be doing promotion for people who understand the importance of communication in the modern world? This is a communications war that the Taliban are waging. They understand that public opinion is probably the Achilles heel of the international community".[39]

Slowly, and often chaotically, the Taliban grew their understanding of the importance of targeted information as a component of their military

---
36 *Videoproduction*, YouTube, available at https://www.youtube.com/watch?v=6RAYjDRJi3s.
37 http://www.youtube.com/watch?v=5jCUeUNbnQE&feature=related.
38 http://www.indepedent.co.uk/news/world/europe/paris-match-taliban-photoshoot-shocksfrance-919109.html.
39 Lichfield J, '*Paris Match Taliban Photoshoot Shocks France*', Independent, 4 September 2008.

campaign. They may have sought to regress their country by many hundreds of years, but this didn't mean they could not apply an agility of mind and, perhaps more astonishing, a highly developed grasp of the role of information to their heavily outgunned and outnumbered insurgency. Indeed, the words of Mullah Abdul Salaam Zaeef, a former Taliban ambassador to Pakistan, are particularly interesting. Holding his iPhone close by he provided an interview to the Associated Press at his residence in Kabul, Afghanistan on 25 February 2009. Zaeef had spent almost four years in a Guantánamo Bay prison. Wearing a black turban and resplendent in thick beard, he admitted to being a big fan of Apple's iPhone. "It's easy and modern and I love it", he told journalists, while he pinched and pulled his fingers across the device's touchscreen to show off photos: "I'm using the Internet with it. Sometimes I use it for the GPS to find locations."[40]

A returning British Commander noted from his time in Helmand: "There has been a lot of talk about asymmetry. The true asymmetry of the campaign is that the Taliban rely on 90 per cent psychology and 10 per cent force whereas we rely on 90 per cent force and 10 per cent psychology in an environment where perception is reality, memories are very long and enemies easily made", prompting a visiting researcher at the Defence Academy to record that: "We are faced by a second generation asymmetric insurgency that is backed by a sophisticated media operation ... that reinforces a number of key but simple messages. These include the need to do duty through jihad, to fight Zionist and Christian aggression that are targeting Islam and setting this in some fourteenth-century period of world history".[41]

In May 2011, I was asked to return to Kabul by a very senior officer running ISAF communication and outreach activities (General H). An enormously experienced, thoughtful, and pleasant man, General H was concerned about some of the proposals that were being submitted into ISAF by commercial companies and I was asked to go to Kabul and review one in particular prior to briefing the then Director of Operations, a US General. The proposal had been submitted by the in-country representative of ICOS (The international Council on Security and Development). ICOS were proposing a project they called 'The Marriage Bureau'.[42] Their proposal

---

40 Geens, S, *Taliban 2.0*, 4 March 2009, available at https://ogleearth.com/2009/03/taliban-20/.
41 Sloggett, D, *IO: The Challenges of Second Generation Insurgencies*, IO Sphere, Journal, Magazine, 2007.
42 See www.IOFFC.info where the full ICOS proposal is provided.

was to provide marriage allowances to young men in selected districts of southern Afghanistan in order to "interfering [sic] with the recruitment of young Afghan men to the Taliban insurgency and reintegrating former combatants into the local communities". In their explanatory note ICOS explained that wedding dowry payments, *walwar*, ranged from $6-10,000 and were important in Pashtu culture as they reflected the social and material standing of the families. As the study rightly pointed out, "these costs are punishingly high in a country where the average annual wage for a man is just $1,825 per year". The basis of ICOS' proposal was that encouraging marriage would have a positive effect on young men, calming them and dissuading them from joining the Taliban. In their justification their listed the results from two sets of focus groups they had canvassed in Helmand and Kandahar where, overall, 40 per cent of the respondents felt this was a good idea. Notably however only 9 per cent of respondents thought that this would stop people joining the Taliban. General H told me he thought the plan was crazy but had been unable to make any progress in stopping its progression through the ISAF headquarters and it had ended up on the Director of Operation's desk. He hoped my presence, as an independent subject matter expert, would help stop it and allow the money to be spent on more useful IO programmes.

In assessing the proposal, I spoke with Human Terrain Specialists in Wardak and Kandahar provinces and asked how many captured insurgents were married: they assessed around half. Contact was also made with British interrogators and translators who told me that at that time most of the Taliban prisoners they encountered were married. Working with colleagues the scheme was costed out. For a pilot programme of 150 young men it would cost between $900,000-$1.5 million; were it to be extended to all eligible young men in Helmand (which was assessed from UNICEF and ISAF statistics to be around 17,000) it would cost between $103 million-$172 million and were it to be extended across the country (an estimate of 760,000 eligible men) the bill would be between $3.9 billion-$6.6 billion. Aside from the annual costs to the US taxpayer, and the paucity of information to even suggest that marriage was an inhibitor to joining and fighting for the Taliban, seven further points were made in what was to be a four-page briefing presented to the Director of Operations.

Firstly, we assessed that the scheme would create considerable discontent amongst those not selected for a dowry – and particularly in provinces where there was no support for the Taliban. Bad behaviours (joining the Taliban) would apparently be rewarded and we wondered if this might actually trigger support for the insurgency in currently peaceful districts. Secondly, there was a very strong possibility of money being stolen or mis-appropriated by Afghan auditors who would be required to administer it. Thirdly, we worried that the scheme would be abused. How would the money be spent and might it be used to purchase weapons. Next, we worried that the Government and ISAF might be seen to be encouraging the procurement of underage brides – a regular and unpleasant occurrence in Afghan society. We also feared that the scheme, which would have to have an Afghan government face on it, would be interpreted as government interference in family and tribal affairs. We wondered if the scheme might be seen as being un-Islamic (something that ICOS themselves had drawn attention to). Finally, we worried that the scheme would become a massive source of unearned income (creating a so-called *rentier* economy)[43] with all the attendant micro and macro-economic consequences that would bring.

Going into the meeting with the Director we felt confident that we had presented robust and thoughtful answers and that the scheme would be dropped; we also had a series of alternative proposals, specifically one that would look at reasons Afghans were leaving the Afghan Army and Police at alarming rates. However, the Director was unconvinced by our protestations. Some ten years later, as Afghanistan fell to the Taliban, I was reminiscing with the now retired General H and the subject of the ICOS proposal came up. I knew that it had eventually been 'canned' but not in Kabul, I think it been pushed all the way to Washington. H looked at me in bemusement. Did I not know that the Strategy Director was married to a senior figure at ICOS?!

Commanding the UK's Psychological Operations unit in southern Afghanistan revealed to me just how poorly we understood the Afghans on the ground. Our small unit had just 8 members, but we leveraged off a wider network of 15-20 local Afghan employees. They were drawn from the local community; most but not all spoke English – some to a very high standard. The women arrived at the gates to Lashka Gar in Burqas, often on the back of a brother or husband's motorbike, shedding the enveloping blue shroud as soon as they cleared security to often reveal bright dresses and highly educated minds. The men were a mixture of ages and backgrounds – many of them graduates from university. All had passed basic security checks, and all were thoroughly searched on entry. Talking to them was

---

43 Rentier capitalism describes the economic practice of gaining large profits without contributing to society.

always a joy and they were honest and candid in their assessments of life beyond the perimeter fence.

Supporting the PsyOps team was the CJPOTF – a Combined Joint PsyOps Task Force – based in Kabul and largely staffed by a rotating team of US, Romanian and German officers with commercial support services. CJPOTF were responsible for coordinating activities across the various deployed multinational PsyOps units and, through a central budget, helping them with costs such as printing, research, and relationship building items (RBIs). Lots of money was spent on RBIs – as the name suggests they were items that would be given to Afghans to foster a positive relationship. Many were indeed useful; duvets, cooking utensils, cricket sets, warm clothes to name a few. However, there were a great many that seemed poorly thought through. In 2008, for example, the US military dropped soccer balls (footballs) to local villages. Designed as a fun gift these RBI balls displayed flags from countries all over the world. Unfortunately, this included the Saudi Arabian flag which features the shahada, the Muslim declaration of faith including the word 'Allah'. This managed to cause enormous offence and local religious leaders criticised the US forces for their insensitivity. Around 100 people held a demonstration in Khost in protest and Afghan MP, Mirwais Yasini, told the media at the time: "To have a verse of the Koran on something you kick with your foot would be an insult in any Muslim country around the world."[44]

As well as the Saudi flag, many items started to arrive branded with the ISAF logo. This too seemed curious since everything that ISAF did was supposedly in support of the Afghan government and if there was to be a logo, we collectively felt that it should be the Afghan flag. However, even that was problematic since the presence of any logo, ISAF or government, would place the owner in danger from the Taliban. Patience with RBIs was lost when the PsyOps team received a consignment of fluffy yellow ducks – children's toys – with the ISAF logo emblazoned on their chests. It was difficult to understand how such an idea could have been considered suitable for dispersal to Afghan kids and in any event how would a yellow duck – which looked remarkable unlike anything that could be found in Afghanistan – positively aid the counter insurgency efforts.

Other RBIs were the subject of a meeting between ISAF staff and US Embassy Kabul staff which, again, General H asked me to attend. The RBIs

---

44 Leithead A, *'Blasphemous' U.S. Soccer Balls Anger Afghans*, ABC News, 26 August 2007, available at https://abcnews.go.com/International/story?id=3525777.

on this occasion were baseball flyway packs – literally large packages of all the necessary equipment to play baseball. The idea had come about from the teacher training college; here aspirant Afghan teachers undertook their professional training – funded by the international community – before dispersing around Afghanistan to begin their teaching roles. Keen to build teamwork and encourage peaceful activities amongst the local youth, it was suggested that teachers could be issued the baseball packs to take with them when they left Kabul and start running baseball games in the provinces for the national youngsters. The idea seemed to have some merit – the role of sport as a unifier is well documented in many different societies – but no one in ISAF could quite understand why baseball, and not cricket, had been chosen as the preferred sport.

Cricket had been played in Afghanistan since the 1800s – almost certainly a legacy of the British presence in the country. An Afghanistan Cricket Board was formed in 1995 and amazingly, given all the problems of the country, Afghanistan became an affiliate member of the International Cricket Council (ICC) in 2001 and a member of the Asian Cricket Council (ACC) in 2003. To understand how Baseball was chosen, a meeting at the US Embassy was convened with the US sponsors and the job of explaining cricket to the US staff fell to one of the only Brits present – me. Whilst trying to explain the rules of cricket proved a reasonably thankless task, I did manage to impart that Afghanistan had a rich history of cricket; it had almost no history of baseball. Thankfully one of the senior US Embassy staff present agreed and the decision was taken to purchase cricket equipment instead. But this in turn revealed a new problem; the money, it appeared, had been secured from the US Congress on the basis of baseball. To change the request would require some form of congressional approval and the US contracting officer on the staff was not at all confident that would happen easily, or at all. I never did find out if the cricket bats, balls, wickets, and pads were procured but Google reveals quite a few references to Afghan baseball. What it did serve to reinforce was the disconnect between (well meaning) Washington (and London) policy, and reality.

In part that disconnect was fuelled by ignorance of what was really happening on the ground and, in an attempt to understand it more, Afghanistan became one of the most heavily surveyed countries in the world. How useful those polls were is debatable. For example, in 2010 the Asia Foundation published its annual longitudinal poll of the Afghan people.[45]

---

[45] *A Survey of the Afghan People,* The Asia Foundation, available at https://asiafoundation.org/where-we-work/afghanistan/survey/.

In its questions about the Afghan Police[46] the poll found that 84 per cent of the (surveyed) population apparently stated that they either 'strongly agreed' or 'agreed somewhat' that the ANP were honest and fair with the Afghan people. Yet this seemed to fly in the face of the very next question which found that 58 per cent of the (surveyed) population believed the ANP were unprofessional and poorly trained. Regardless of the polling results, for those of us on the ground in Helmand, the behaviour of the ANP was a daily concern. Whilst many were professional there were a significant number who behaved extremely poorly, often involving themselves in extortion, more general criminality, and even predatory sexual behaviour. Although these behaviours caused considerable disquiet, they were very difficult to address – maintaining a partnership with local Afghan security forces was paramount to the military missions and anything that might threaten that was dealt with very cautiously, if at all. Often coalition commanders had to, metaphorically, 'hold their noses' else their comments be taken as offence.

In the PsyOps office in Helmand, a US consulting company – McNeil Technologies – had embedded a contractor, a former US infantry officer, in our HQ; he had responsibility to run their Afghan atmospherics programme which was designed to 'integrate and apply socio-cultural knowledge of the indigenous civilian population by 'passively' gathering information'. They populated the Helmand Perception matrix (HPM) – a grid square map of the province, colour coded (green) to indicate where there were positive perceptions of ISAF and red to indicate problem areas. How these squares turned green or red was always a mystery until I sat down one day with the small US-Afghan team and asked them to take me through it. It transpired that an Afghan would hand out leaflets to people in different locations and ask them to ring a mobile phone number – located inside the base and manned by an Afghan American – and explain their perceptions of the security situation. If they did, they were rewarded with a small number of Afghanis being credited to their mobile account. At the evening command briefs in Task Force Helmand the HPM would be shown; it became an important component in a commander's decision making over where troops would be tasked. The potential lack of rigor in the process was a concern for us all. As we saw in an earlier chapter, people's attitudes and perceptions are poor precursors to behaviours yet here that relationship, weak as it was, was being taken as gospel and informing real

---

46 Ibid. Table 3.6: Public agreement and disagreement with statements about the ANP (Q-36a-e, Base 6467).

life military decision making. I also wondered if the inducement to ring – Afghani credits – resulted in genuine information, from real Afghans, or information provided just to get the reward. Or worse still, deliberately misleading information from Taliban or Taliban supporters.

The presence of advertising billboards and posters across the country also caused disquiet. Across Helmand, and wider Afghanistan, some contracting company had won the contract to construct, print, and erect large billboards with pro Afghan government messaging. In Helmand a number of these were focused on poppy eradication. Quite aside from many, if not most of the intended audience being illiterate, the message, which read "Pashtu. The Poppy is your enemy"[47] clearly had little resonance with an impoverished people who did not regard the poppy as an enemy, rather they viewed it as a (cash) gift from God that, God willing, came every year and which would help pay for vital household essentials. Yet the billboards sprung up everywhere and at vast expense.

Sometimes good ideas were just poorly synchronised. After years of trying to explain why foreign troops were in the country at all, in late 2011 ISAF issued a DVD to its troops on the ground entitled *'Why we are here'*. It provided a script for troops to use which told them that: "This video contains images, sound and text associated with the reasons as to why we (the coalition) are here, what we are doing and when we are leaving". But the timing was problematic as various Afghan governmental ministries were engaged in the first stages of transition – the passage of responsibility for security away from NATO to the ANSF. The issue of the DVD was a welcome, if very late, initiative. In 2011, ten years after the coalition removed the Taliban most Afghans outside of the cities, and probably a large number within, had no idea why foreigners were in their country. As a PBS news article revealed, knowledge of the attack on the twin towers was almost non-existent.[48] However, the DVD presented real problems to troops on the ground who were struggling to simultaneously explain in ways that resonated with their specific local audiences, why ISAF was there and why it was leaving. But hopes were not high; as RAND noted in their paper on US IO: "the average Afghan is not really interested in our 'narratives'".[49]

---

47 See www.IOFFC.info for a picture of this particular billboard.
48 *What Does 9/11 Mean to People in Afghanistan?*, PBS News Hour YouTube Channel, available at https://www.youtube.com/watch?v=SimIS_cQ6ko.
49 Munoz A, *U.S. Military IO in Afghanistan Effectiveness of Psychological Operations 2001–2010*, Rand, available at https://www.rand.org/content/dam/rand/pubs/monographs/2012/RAND_MG1060.pdf.

This is not to say that all IO in Afghanistan were a failure. When focussed directly on behaviours – as opposed to attitude and perceptions – there were notable successes and perhaps the most important of those was during the various elections when IO was used to encourage people across the country to register as voters and then vote in parliamentary elections. This was by no mean feat; unless you have been to Afghanistan and experienced the totality of the nation you will not be able to understand the logistical, communication, security, and physical difficulties of running an election. From the very start people realised that any election in Afghanistan would not be comparable to one in Europe or the US; it had to be 'good enough for Afghans' but everyone recognised that there would be massive challenges. And so did the people and the prospective parliamentarians who in many instances would risk their own personal safety (especially the women) in declaring their intent to run for office. From a slow start in 2004, just over 40 per cent of the population voted in the 2010 elections. Although the 2010 vote was marred by some violence and loss of lives, 41 women were selected as MPs and over 4 million people voted. IO played a significant part in informing people of the process and reassuring them of their security and safety. People also needed to know who they were voting for, particularly with such crushing rates of illiteracy. Contrary to some of the more conspiracist laden voices online, western forces didn't set out to secure one party or an individual's election over another. That was not the point of the IO campaign; IO was used to help get the electorate out and to inform them of their choices and how to vote. IO was not there to tell people who to vote for. In that respect the cross-country IO, largely conducted by radio broadcast, was successful.

Radio was a hugely important means of messaging the population and the small team from 15 (UK) PsyOps Group, based in the small town of Lashkar Gah, spent their days talking with and listening to the Helmand population. Principally that was done via the eight local radio stations, broadcasting around the clock, that the group had established and maintained. Despite the military might of the US, the UK radio stations were recognised as examples of best practice and 15 (UK) PsyOps's broadcasts were retransmitted by a further 16 US radio stations giving the group reach across the whole of Helmand and even into parts of neighbouring Kandahar province. The radio stations provided local, provincial, national, and international news, connecting the population to world events, facilitating informed discussion and promoted a sense of community and national identity. Regular phone-in programs allowed the population of Helmand to freely express their opinions from the security of their own homes. In doing so, communities were reconnected and civil society, previously so susceptible to insurgent intimidation, gradually built and, importantly, those who were previously disenfranchised are given a voice; that voice may criticise the Taliban or indeed ISAF and the Afghan Government; but importantly there was no censorship.

Many of 15 (UK) PsyOps's radio programmes had come about as a result of the population's own requests, including programmes on women's health issues. As well as an extremely popular sitcom based upon the long-running BBC radio drama 'The Archers' (which interestingly the BBC started post-World War II to re-teach good farming practices, after the loss of so many farmers and labourers) called *Chai Dawat* (*The Tea Shop*). Poetry and jokes were also staples for phone-in requests and revealed a rich vein of humour and culture – humanity – that in a conflict environment would never ordinarily be visible. Local governmental representatives were encouraged to appear on programmes allowing them to be held responsible for their actions and in turn to educate and inform their constituents. This was particularly important during the Lashkar Gah Municipal Elections when the radio advertised the elections existence and explained to the population how they could register to vote and participate, and who the various political contenders were. All of this is a mainstream and everyday feature of western electoral processes but in Afghanistan this was still very much a new process and one that required careful and sympathetic education and tutoring.

Helmand is a subsistence agriculture economy; the practices of sowing and harvesting have changed little in hundreds of years. 15 (UK) PsyOps radio stations, and its people 'out on the ground' helped promote the distribution of wheat seed as an alternative to poppy production. When British Army vets visited communities to run clinics and teach animal husbandry, 15 (UK) PsyOps was invariably present, recording their surgeries and lectures, and rebroadcasting them across the province to growing and receptive audiences. 15 (UK) PsyOps had advanced design and production capabilities; when combined with the group's deep understanding of the communities and population they designed and produced suitable training and pictorial educational aids, informational posters, and radio advertisements to assist the veterinary programme and, in particular, to alert farmers to the dangers of IEDs.

Of course, there was also a harder edge to UK PsyOps; in some situations, military force was the only solution to a particular problem.

The British TV presenter, Ross Kemp, in his 'Extreme World' TV series spent time with some of the PsyOps forward units and saw how radios and loud speakers were used to goad hard core Taliban into exposing their positions.[50] PsyOps were also used to support Special Forces operations, particularly the targeting of high value Taliban commanders or bomb-makers, although this was never particularly successful in Helmand as UK Special Forces were generally wary of involving anyone outside their immediate community in operations.

Ingenuity was always on display in the team. None more so than the design and production of the LRGR by 15 (UK) PsyOps. A local Patrol Base commander was worried that children kept throwing stones at his position; they had no way of knowing if the stones were in fact grenades, and he was concerned that a child may get shot by accident either by a British or ANSF soldier. Various methods had been used to try and stop the children, but none had proven successful. The 15 (UK) PsyOps team set about designing a solution using a length of drainpipe, an axel stand, a flashing red LED, a pretend aerial and all held together with black masking tape.[51] Christened the LRGR, the Afghan police were briefed on the 'secret rays' that it emitted and how, if you stood in front of it, your ran a very real danger that your reproductive organs would be irradiated; LRGR was the abbreviation for the Long-Range Gonad Reducer. The LRGRs would be placed at either corner of the Patrol Base and pointed outwards. Relying on the Afghan Police's inability to keep a secret, and every Afghan male's 'alpha male / macho' outlook on life, it was hoped that the bizarre looking device with its flashing red light would deter future stone throwing. LRGR was completely successful.

Another issue 15 (UK) PsyOps had to deal with were the so called 'Night Letters'. These were written notes, from the Taliban, pinned to compound doors overnight, threatening the compound occupants, be it for interacting with ISAF troops or not supporting the Taliban vocally enough. These were extremely disconcerting for the population and they felt that at night, in particular, they were particularly vulnerable to intimidation and reprisals. In response UK and Afghan patrols were issued with stencils and white powder sprays; the stencils guided the white powder onto doors to spell out that ANA/ANP had patrolled that night and were looking

---

50 *Man Down on the Rooftop*, Ross Kemp Extreme World, YouTube channel, available at https://www.youtube.com/watch?v=zJ0bchTPsto.
51 See www.IOFFC.info for a picture of the LRGR.

after their security. The white powder could then be scrubbed off the next morning. With the odd exception, the IO battles in Afghanistan were largely local affairs. US Airforce Major, Frank Lazzara, wrote in his 2012 War College dissertation: "[IO] hasn't been all failure in Afghanistan, but success seems to be limited to tactical and regional instead of central and operational".[52]

Regardless of the tactical successes, the humiliation of August 2021 is hard for all Afghan veterans, particularly those who worked alongside Afghan personnel whose fate is unknown. Evanna Hu, a non-resident senior fellow at the Atlantic Council's Scowcroft Center for Strategy and Security, wrote:

> "A large portion of Taliban-conquered territories saw little to no fighting, thanks to the group's strategically mounted psychological operations toward the Afghan military and high-level government officials, as well as the Afghan population. When targeting the former, the Taliban have made political promises and inflated the number of troops they have. They have also formed a narrative that the Afghan government's allies, chiefly the United States, have abandoned them, in addition to highlighting the government's wrongdoings over the past two decades, particularly widespread corruption. To the population, meanwhile, the group is promising stability and communicating the fact that their lives under the government aren't any better. Coupled with the psychological-cultural role of the Taliban as a "boogeyman," the group has successfully fashioned itself into a larger-than-life threat."[53]

All of that is true and the command of ISAF did not understand the human dimension of the conflict and in particular what motivated behaviours in Afghans. Even when presented with the tools to do so, they never saw their value. What little research was commissioned was always a poll of some kind. Anecdotally, Afghanistan must have become one of

---

52 Lazzara F, *IO, Finding Success as Afghanistan Draws to a Close*, The Naval War College, 4 May 2012, available at https://apps.dtic.mil/sti/pdfs/ADA563888.pdf.
53 Hu E, *The secret to the Taliban's success: PsyOps*, The Atlantic Council 15 August 2021, available at https://www.atlanticcouncil.org/blogs/new-atlanticist/experts-react-the-taliban-has-taken-kabul-now-what/#hu.

the most polled nations on earth. Yet the polls were at best a temporary snapshot of an attitude; at worst they were a 'guess' but the degree of importance attached to them by senior commanders was astonishing. One of the great 'gurus' of political polling in the US is Frank Luntz. He had been critical of the polling industry in 2016 which suggested that Hillary Clinton would get easy wins over Donald Trump in certain states. In the wake of the 2020 US election, Luntz declared that "The political polling profession is done"[54] and The Washington Post declared that: "We still don't know much about this election — except that the media and pollsters blew it again".[55] If the US political system, with its billions of dollars and its Political Action Committees has decided polling has very limited use in US elections, surely the time has come to banish it from the military's tool kit? There are many lessons to be learnt from Afghanistan, but my hope is that proper behavioural research will be undertaken in future conflicts to guide senior military officers' decision making; and never again must perceptions and attitudes (and their determination through polling) be used in the way they were in Afghanistan and Iraq.

---

54 Concha J, *Frank Luntz: Polling profession 'done' after election misses: 'Devastating to my industry'*, The Hill, 4 November 2020, available at https://thehill.com/homenews/media/524478-frank-luntz-polling-profession-done-after-election-misses-devastating-to-my.
55 Sullivan, M, *We Still Don't Know Much About this Election*, The Washington Post, 11 Apr 2020, available at https://www.washingtonpost.com/lifestyle/media/we-still-dont-know-much-about-this-election--except-that-the-media-and-pollsters-blew-it-again/2020/11/04/40c0d416-1e4a-11eb-b532 05c751cd5dc2_story.html.

5 Understanding Audiences

In the north of England mushy peas are a traditional delicacy. Mushy peas are dried marrowfat peas, soaked overnight in water and baking soda, rinsed, covered in water and brought to the boil until the peas are softened and mushy. As delicacies go it does not look the nicest – a green glutinous mass on the side of your plate – but it is none the less, popular. I often used to set my military IO students a task to design a programme to influence people to buy more mushy peas.

Most students immediately went down an advertising type route. I had suggestions that perhaps highlighting the healthy credentials of vegetables might work.; some suggested TV adverts or maybe endorsements by celebrities ('David Beckham eats mushy peas'!). For years I used to show them a picture of Chalfont Viaduct in the UK, a bridge spanning the busy M25 London orbital motorway, which up until 2018 had the words 'GIVE PEAS A CHANCE'[1] graffitied onto the side. I would joke with them that this was either the work of a dyslexic peace protestor or it was part of some past IO student's Mushy Peas marketing campaign. But jokes aside, the prospective sale of mushy peas provides a helpful example not of advertising techniques but much more importantly, how we need to understanding audiences if we are to have any hope of influencing their behaviours.

The first thing we need to know is, who is the audience? This is a simple question but oddly one that commanders and policy makers struggle with. In the 2003 Iraq War there were two very different audiences; the domestic UK and US audience (with whom support for the troops would be vital) and, secondly, the Arab world and particularly those who may be drawn to support or finance the insurgency. Two different audiences requiring often very different messages. In the case of the Mushy Peas campaign let us assume the audience is exclusively the British public. Why? Experience from my multinational IO classes shows that when a PowerPoint slide of mushy peas is shown, Canadian and American students will presume its

---

1 A picture of the graffiti is provided at the book's website, www.IOFFC.info.

guacamole and mainland Europeans will screw up their faces that whatever it is, it looks horrible.

The next stage is to ask, under what circumstances would the audience – Brits – consider eating mushy peas? And here the most detailed understanding of the audience is key to deriving the solution to the problem. Because if you understand Brits then you will know that (probably) in 95 times out of 100, Brits will eat mushy peas under only very specific circumstances. It is often joked that the UK's national dish is Fish and Chips (French fries) and indeed it is enormously popular. None more so on a Friday lunch time because under Christian teaching Jesus died on a Friday and a way to honour his sacrifice was to abstain from meat. Fish though, as a cold-blooded creature, was considered acceptable and so the Fish on Fridays tradition began.[2] The accompanying mushy peas, which began as a north of England delicacy, have now extended south but probably not much further from the UK's borders.

We now know three important additional pieces of information. The behaviour – eating mushy peas – is largely confined to Brits, it is largely undertaken on Fridays and the vast bulk are consumed when accompanying Fish and Chips. This is quite important research – in the language of Psychological Operations we refer to it as Target Audience Analysis (TAA) – because it leads us to an obvious solution to our problem; to increase sales of mushy peas, invest no time at all in their advertising. They might be tasty, but they look unpleasant and no celebrity is ever going to put their name to them. Instead spend all your time promoting Fish and Chips, because by implication consumption of mushy peas will increase. The problem of presenting a product that looks and sounds so unappealing is overcome and the end effect, increased sales of mushy peas, is achieved. But to most advertisers the idea of taking the product out of the advert is anathema. This rather silly story serves to illustrate to IO personnel that the solution to the problem is always to be found in the underpinning research of the intended audience, is rarely what the 'client' envisages the campaign will be and is often highly counter-intuitive.

Chris Donnelly was the principal advisor to the Secretary General of NATO for many years. He was a visionary in the strategic intelligence community and his areas of interest and expertise ranged widely. One of his enduring concerns was that defence would struggle to keep up with

---

2 *Reasons why British people east fish and chips on a Friday*, available at https://chefstravels.com/why-do-we-eat-fish-and-chips-on-a-friday/.

the pace of technology and research in the commercial world. After leaving NATO and joining the MoD, he established a small internal think tank to advise and assist the service chiefs. The Advanced Research and Assessment Group (ARAG) based at the UK Defence Academy and co-located with the Conflict Studies Research Centre, was home to most of the MoD's Russian linguists. Chris asked me to join his team as Director of Communication's Research and one of my first tasks was to evaluate what the commercial communication industry had that might benefit UK defence, which at that time was struggling in both Iraq and Afghanistan with sophisticated insurgencies. Together with other colleagues, including a seconded senior Defence Science Technology Laboratories (DSTL) scientist, we invited a host of communication companies to present their offerings and research. Most of the 'big' UK names presented but so too some of the smaller and less well-known companies.

One of those was SCL Defence, a UK company that had existed since the early 1990s but was making its first foray into defence. It had a larger parent company – SCL Group – with a previous track record of political lobbying and electoral communications but SCL Defence was a new offshoot with the UK Security Clearances (so called List X) required for defence work and therefore heavily ring fenced off from the rest of the group's business.

SCL Defence were the standout company because they articulated two very clear points that resonated from operational experience. Firstly, they told us that UK armed forces did not understand why people behaved in the way they did and because of that simplistic advertising-based communication programmes were used which owed any success they may have had more to luck than science (and in particular, the Social Sciences). Secondly, they focussed on behaviour, which they said was much more important than attitudes, and their research always began with one simple question: 'Under what circumstances would an audience do, or not do, something?' To answer this latter question, they told us that internet scaping – which many communication's companies were advocating – was probably next to useless and instead they used field-based research to get to the nub of behavioural problems. They presented some fascinating past case studies.

In the Caribbean they had been involved with CARICOM – the intergovernmental organisation of 15 Caribbean member states – in trying to stem the transmissions rates of HIV/AIDs. Officials were alarmed that if rates were not reduced within 25 years, the populations would be decimated. Money was secured for a campaign to encourage men to wear condoms. Yet this campaign met significant resistance from its target audience. Caribbean men did not wish to wear condoms; in their mindset, the wearing of a condom was counter to their masculinity; and you only wore a condom if you had the disease. SCL showed us how, through qualitative and quantitative research, they had switched the target audience from men to women and the messaging from countering HIV to countering illegitimate births. The latter was a significant cultural taboo and unmarried mothers were in many instances outcasts from their communities. Whilst the end effect was the same the change of audience and message was significant.

In Saudi Arabia, in the wake of the 9/11 attacks, the Kingdom sought to distance itself from some of its more extreme preachers. A concerted effort was made to find more moderate teachers and preachers, with a new educational syllabus introduced. SCL were asked by Saudi authorities to review it and found that whilst young Saudis would always be deferential to authority figures – Imams and teachers in particular – more credible conduits for messaging came from unlikely sources. Premier League football players were found to be particularly authoritative figures and counter-radicalisation programmes with well-known sports figures as their sponsors were proposed.

This type of PopInt was available nowhere else and it was of particular interest given the difficulties that coalition forces were experiencing in Iraq and Afghanistan, despite the millions of dollars being spent on strategic communication campaigns. SCL were a breath of fresh air and we looked carefully at their methodology – which they called the Behavioural Dynamics (BD) approach. In the BD methodology they split down the understand process into three phases. The first, which they called descriptive, looked at the background issues to the problem. What was the language used; how literate were people; what groups existed; who were their leaders; what historical context lay behind events; what needs did people in groups have.

In the second phase, which they called prognostic, they looked at issue such as a group's propensity to change its behaviour, they looked at what bound groups together – their affiliations, how strong those were and Locus of Control (which we saw in an earlier chapter). They applied a type of thinking called EMIC logic to this research – a process that tried to think through the problem from the perspective of the people in the group and not from the perspective of outsiders looking in (known as ETIC knowledge).³ This approach, again, was a revelation. We had all tired of communication companies telling us that this message or that message would solve the problem when really they were presenting their own opinions on what should be done drawn largely from their creative backgrounds. We were all very aware of the excellent quote from anthropologist Paul Bohannan who had written that: "There is no more complete way to misunderstand a foreign civilization than to see it in terms of one's own civilization",⁴ a view often expressed in the term ethnocentrism (the belief that the people, customs, and traditions of one's own country and race are better than those of others). We also learned of the perils of homophily – the attraction we have to people who are like us. The final phase of the BD methodology was the transformative. What would messages say, look like, how would they be transmitted? Who would initiate them? And critically how would we measure their success?⁵

All of this was revolutionary, at least for the military if not for the commercial world. Nic Hammerling, former Head of Diversity at Pearn Kandola LLP, told BBC Radio 4: "human beings are very social animals and one of the things that drives [behaviour] is social identity and which groups we belong to." But in the military it looked so unlike anything we had seen before in our IO. It placed the audience and the effect at the very center of what we were doing. It was a solid evidence-based research process. Whilst our CULADS and the US Human Terrain Teams went some way to addressing bits of this, no organized process existed. We collectively felt that this was a genuinely new approach and we urged the wider IO community to look at it closely.

All of this contrasted dramatically with the way that Target Audience Analysis was conducted at the time. Fairly old-fashioned analysis tools were used such as PEMISI and SCAME analysis.⁶ Whilst some training courses for military analysts spoke, briefly, of psychological factors it was rudimentary and basic. But the critical problem facing military analysis was that most of it was done remotely – looking in at groups rather than from inside the group.

---

3 A very useful paper on these two approaches is provided at: https://scholars.sil.org/thomas_n_headland/controversies/emic_etic/introduction.
4 Bohannan, Paul, *Social Anthropology* (New York: Holt, Rhinehart & Winston), 1963.
5 These three phases are shown in a graphic available at the book's website, www.IOFFC.info.
6 PEMISI is Political, Military, Economic, Social, Information, Infrastructure. SCAME is Source, Content, Audience, Media, and Effects.

When I served in the MoD's Advanced Research Group, I wrote definitions for three tiers of TAA which were subsequently incorporated into UK doctrine. Tier 3 TAA was by far the most common, but it was also the least detailed. It was invariably desk and internet based, almost always in the English language and almost always the result of assessing other people's views and opinions. In the definition I called it 'assumed information'.

Tier 2 TAA was any primary research involving contact with audiences, but which does not follow a scientifically deductive methodology. It may be conducted in country or remotely and was largely *attitudinally* based. The output of Tier 2 TAA we defined as information recorded from interactions with target audiences. This was the work of HTS and CULADS and the collation of reports from patrols that went out and interacted with locals. Useful information, and far better than Tier 3, but still a long way from the type of data that was needed to conduct the most effective IO.

Finally, Tier 1 was the name we gave to the SCL approach. This was a multi-source, scientifically verified, diagnostic methodology undertaken in country and in host language used to identify specific motivations for *behaviour*. The output of Tier 1 TAA is information deduced from methodically gathered data and tested against a scientifically derived hypothesis. This Tier 1 material was absolute gold dust and we quickly set about operationalising it. In tandem with the UK's DSTL research organisation, a trial of the BD approach was initiated. In its classified report on Project Duco, as the nearly year-long trial was named, DSTL told both the MoD and SCL that the process was rigorous and opined that if undertaken properly it would work well.[7,8]

The BD process has subsequently been used in many real operations. For example, in 2017 a western government contracted SCL Defence to undertake a review of their global counter ISIS messaging globally. Across some of the most dangerous countries in the world researchers undertook field research, conducting 160 qualitative interviews and 240 quantitative interviews. Researchers spoke to 11 active ISIS recruiters, 108 potential ISIS recruits and 41 returned fighters. Using the three phases of research, SCL Defence provided a report back to the government client that showed that the

---

7 Detailed debrief provided to Nigel Oakes of SCL, on cessation of the project, and relayed to the author, by Oakes, immediately afterwards.
8 A highly redacted version of this report (DSTL/CR79142 1.0), with almost everything removed, is available at https://www.whatdotheyknow.com/request/389795/response/975478/attach/3/FOI%202017%2003434%2020170508%20Rpt.pdf.

top four reasons for the recruits joining ISIS were very similar to the reasons young men and women in the client nation joined their national armed forces: money; excitement; empowerment; a better life. What really surprised the client was that religion was not found to be the primary driver for ISIS recruits. In the presentation to the nation, SCL Defence made the following recommendation: "The religious influence should be treated as contextual framing, in much the same way as national pride. Base motivations are likely to be much more effective than higher needs in motivating recruits".[9]

This level of understanding is unprecedented; up until this moment opinion polling and conventional military intelligence gathering had been the primary means of determining communication strategy and was self-evidently not working well. With this level of potential understanding there was now a very real chance of designing persuasive, behaviourally focussed, communication programmes that would deter violent behaviour and terrorist recruiting. It was quite literally a watershed moment but one short-lived. Within just a few days a story about a separate part of the SCL Group called Cambridge Analytica would break and within a few weeks the SCL Group had collapsed under the weight of an unprecedented global media scrutiny in the wake of Donald Trump's election as US President and the UK's Brexit referendum. Nevertheless, SCL's TAA had firmly established its credentials, at least in the US where it chimed with long held beliefs about US influence and outreach.

Professor Richard Cottam was Professor of Political Science at Pittsburgh University in 1969. His area of interest was political interference, a form of diplomacy which involved counterinsurgency, political, economic, and psychological intervention. In his book on the subject, Cottam wrote that: "our diplomats and policy makers have been trained in old style diplomacy and still adhere to its principles and speak its language. However, the discrepancy between understanding and practice leads to a hit or miss kind of foreign policy lacking in long range planning and theoretical coherence". The remedy, in his view, was something he called 'Competitive Interference'.[10] In 2013, the term was resurrected when Nadia Schadlow, a former member of the US DOD's Defense Policy board, wrote *"Competitive Engagement: Upgrading America's Influence,"*[11] and argued that

---

9 Final presentation of findings of SCL Defence findings to client. 12-14 March 2018.
10 Cottam, R, *Competitive Interference and Twentieth Century Diplomacy*, 1967, University of Pittsburgh Press.
11 Schadlow N, *Competitive Engagement: Upgrading America's Influence*, Orbis Volume 57, Issue 4, Autumn 2013, Pages 501-515, available at https://www.sciencedirect.com/science/article/abs/pii/S0030438713000446.

organizations across the US government that work overseas have to think about the challenge of their operating environments in ways that deal with the competitive nature of those interactions: "Being successful in a competition requires knowing and understanding both one's competitors and oneself" she declared. In the same article, Michael P. Noonan, the Director of the Program on National Security at the Foreign Policy Research Institute in the US, wrote: "Clearly, the president needs options between military intervention and complete non-intervention ways to influence developments in the Middle East without deploying Reaper drones or sending U.S. ground forces. To give Obama the tools he needs, the U.S. government should reinvigorate its capacity to wage 'political warfare.'" Defined in 1948 as 'the employment of all the means at a nation's command, short of war, to achieve its national objectives', political warfare uses both overt and covert measures and can range from political alliances, economic measures and 'white' propaganda to covert operations such as clandestine support of 'friendly' foreign elements, 'black' psychological warfare and even encouragement of underground resistance in hostile states."[12]

Yet to do so today would be extremely difficult for two specific reasons. The first is that it is increasingly difficult to keep operations covert. Social media watchers can be remarkably quick in spotting untoward events – the mission to kill Osama Bin Laden was very quickly made public by IT Consultant Sohaib Athar, who was watching events outside his house and live posting on social media.[13] Alongside social media are Freedom of Information requests – the seeming ability of any member of the public to demand access to specific government programmes or projects. Both have served to make IO, in particular, very difficult. And then there are the social media companies who stung by a liturgy of criticism are very slowly tightening up who can (and who cannot) use their platforms. But by far the bigger inhibitor to successful operation is poor understanding of the problem. For all the 'Ints' that we have in our collective armoury we are still missing the 'PopInt' of the very advanced Target Audience Analysis type that SCL developed, and that from a short period of time the MoD and NATO used to great effect before it disappeared under the weight of the Cambridge Analytica scandal.

---

12 Ibid.
13 https://edition.cnn.com/2016/01/20/asia/osama-bin-laden-raid-tweeter-sohaib-athar-rewind/index.html.

6 Past IO Campaigns

Examples of good and bad IO are provided throughout this book. But some operations are either of such a magnitude, are of such complexity or alternatively are so simple, that they deserve special mention. I have drawn on nine examples. From Jamaica and from the Philippines, two examples of operations that involved a whole of government approach, significant coordination and were devised to deal with very specific and large-scale issues. From Colombia, the simplest of ideas provided the genesis of an incredibly ingenious campaign; in Syria the Islamic State, known for its brutal terror videos, ran the simplest and subtlest of IO that led to the defeat of the Iraqi Army in Mosul. In Ukraine, Russian integrated cyber and IO to engineer localised chaos and Ukrainian ingenuity used to identify Russian War criminals. In Israel/Palestine, two examples of the difficulty of using information to win opinion when it is coupled with over-whelming force and civilian casualties and finally, from my own experiences, the broadcasting of messages from ships into Iraq in the 2003 war.

Jamaica

In 2003 it had become clear that Jamaica had become a major hub for drugs trafficking in the Caribbean – with much of it finding its way to the streets of the US and Europe. The island itself was in the midst of a significant crime wave, with a rising murder rate and a huge growth in gang activity; the two were clearly connected and the gang leaders, known locally as Dons, Strong Men or Area Leaders, were engaged in a massive fight for control of a potentially multi-million-dollar market. In part they had been allowed to grow and even flourish because of the long-standing political and police corruption which in turn had left the population scared and unwilling to work with authorities.

Clearly the situation could not be allowed to continue, and a group of officials met to consider how the problem might be tackled. A range of Jamaican open-source materials were studied, in particular the websites of the Jamaican national newspapers: *The Gleaner*, *The Observer* and *The Star*. The letters pages and the Op-Ed commentary columns were particularly useful in understanding the views of the Jamaican general public. Research revealed that Jamaicans were very proactive with phone-in radio talk shows and a huge number had grown to meet demand. They were all listened to carefully to help form a view of everyday life on the island. Polling revealed very negative perceptions toward the security forces due to corruption and lack of trust, with between 40-50 per cent of respondents saying that they would *not* contact the Security Forces, either because they feared for their own safety or because they quite simply had no trust in the police. An additional and important finding was that Jamaican society (which was strongly matriarchal, church-going, and homophobic) had a well-developed underlying anti-informant culture.

With a clearer understanding of both problem and the audiences, a specially formed team settled on the development of an IO-based strategy in support of specific activities by the Jamaican authorities and in particular a brand-new anti-crime Task Force, which would become known as Operation KINGFISH. That task force, which would be staffed by trusted and vetted members of the Jamaican Constabulary Force (JCF) and the Jamaican Defence Force (JDF), would be augmented by experienced officers from other nations who would help develop an IO strategy that would inform and ideally 'sell' to the Jamaican population the new Task Force and establish it as being enduring, credible and honest. It also needed to have clear and achievable aims.

Officials began looking at how the Jamaican audiences could be reached, and several problems were quickly identified which meant traditional methods would likely not be practical. Although mobile phones were very common, mass SMS text messaging to the population to encourage them to ring the new confidential tip line (based, for operational security, 'off-Island' in a third-party country), was quickly discounted because of a curious anomaly that in Jamaica mobile users pay to receive texts rather than pay to send. Billboards could not be used because a recent hurricane had destroyed most of the islands advertising space and airborne leaflet drops over the areas under the control of the Dons were discounted because, culturally, littering in Jamaica is very much frowned upon.

Relationship building items (RBIs) such as wristbands, ball caps and badges could not be distributed in case anyone wearing them become a target for the gangs and using local influencers would also be problematic as some of the key personalities, such as popular Reggae artists, was discounted because many of them were known marijuana users which would run counter to the implicit anti-drugs message. This latter issue was a significant disappointment as there were already case studies of how music and musical influencers could assist IO campaigns. For example, in Sierra Leone a group of local/regional artists had been encouraged to write and record the best dance track they had ever performed and were provided the influence themes on which to base their lyrics. The song 'Salone Soja' went viral and for months the population were singing and dancing to the track which the newly trained Republic of Sierra Leone Armed Forces (RSLAF) took as their anthem.[1]

The Task Force had to be shown to be different to previous, failed, initiatives. To achieve this some of the key messages revolved around the honesty and integrity of its personnel, who were supported by international partners and that the Task Force's work would be intelligence-led. To facilitate that, the anonymous telephone tip-line was established and advertised by a series of hard-hitting TV and radio adverts. With the support of the then Minister of National Security, Dr Peter Phillips, the Task Force was stood up and three key objectives were established; persuade the population to support Operation KINGFISH; convince the population that the Dons were not 'untouchable' and convince the Dons and gangs that their days were numbered. These local objectives happily matched the objectives of European and US governments (who wished to stop or at least dent the quantity of drugs finding their way to their nations) and Jamaica (the removal of guns and a reduction in the murder rate).

As the public launch approached it was felt that a KINGFISH 'quick win' would be useful and in the 48 hours prior to the launch one of the most notorious Dons was arrested on double murder charges, eventually being sentenced to life imprisonment in April 2005. The official launch of KINGFISH was held in the conference facilities of a Kingston Hotel with the Minister of National Security flanked by Assistant Commissioner Hinds, the head of KINGFISH, the two heads of the JCF and JDF and, importantly, by the project's guarantors – US and European diplomats – present. The press conference was followed by a flurry of meetings with media, church leaders and private sector non-government organisations, and in the evening the Minister made an all-station broadcast, personally introducing KINGFISH to the nation. KINGFISH could not come quick enough; on the

---

[1] See *Steady Bongo – S L F A Salone Soja* available online at https://www.youtube.com/watch?v=PZ8NLsdTRYM.

day of the announcement the week's murder rate figures were released; 44 people had died raising the year-to-date murder toll to 1,141.

Quickly an informational drumbeat was established, every activity and success against the Dons and their gangs quickly and widely publicised. Some previously captured 'go-fast' speed boats, used by the gangs for narco-smuggling from South America, were destroyed in front of the media. A known illegal airstrip was destroyed – again with active media attention. Away from the deliberate orchestrated public gaze the serious business of intelligence-led operations began to be ramped up and ten days after the launch, the confidential and anonymous 811-telephone tip-line was launched, initially to mixed results as children began calling because it was free. However, the childish novelty soon wore off and calls from the population started to come in.

Unexpectedly, but to the very pleasant surprise of the IO team, a local Reggae star, Richie Spice, decided to release a single called 'Operation Kingfish' which broadly, and unbidden, mirrored many of the KINGFISH IO messages.[2] The single reached number one in the charts and remained there for a number of weeks, helped behind the scenes by KINGFISH staff who encouraged and cajoled local radio stations to play it as often as possible. Spice's single coincided with a 48-hour curfew imposed in the August Town district of Kingston. Despite complaints from the opposition parties demanding that the curfew was illegal, members of the population began calling into various radio stations demanding that the curfew remain and, importantly, that KINGFISH officers should be allowed to get on with their job. KINGFISH was starting to be effective.

At KINGFISH's launch, Dr Phillips had apocalyptically told the Jamaican population that Operation KINGFISH was quite possibly Jamaica's last chance before it failed as a functioning state and tipped over to a narco-state. The IO campaign amplified Phillips words and began to turn the anti-informant sub-culture against the Dons and to support KINGFISH and also began deliberately targeting the drug traffickers, gangs and Dons to make them uncomfortable and sow division and suspicion amongst them. Of course, the influence strategy could not win alone and the key to success would be the cohesion of the team's various elements, not least the judiciary who succeeded in an almost 100 per cent prosecution rate of KINGFISH targets, all further amplified by the IO team and its growing chorus of advocates.

---

2 *Richie Spice – Operation Kingfish*, The ReggaeChannell available at https://www.youtube.com/watch?v=RVrR2y3pTUw.

A year after launch, the PSOJ (the influential Private Sector Organisation of Jamaica), produced a report that stated that KINGFISH had received a level of public support not seen on previous initiatives; and the 811-telephone tip-line number had taken over as *the* number to call; that 'informant' was no longer a dirty word; that KINGFISH officers (and not the Dons) were perceived to be 'untouchable'. Thirteen months after the launch, *The Gleaner* reported "We urge citizens to continue to avail themselves of the facilities at their disposal through Operation KINGFISH to report known crime or suspicious activity. This is not just a call to civic responsibility but an exercise in self-preservation."[3] The influence strategy was working. Three years on from launch the KINGFISH effect continued to endure. At the time *The Jamaica Star* newspaper had been extremely sceptical about the KINGFISH initiative. But in 2007 they wrote: "Operation KINGFISH has received more than 2,000 actionable calls and has mounted more than 2,000 operations leading to the recovery of nearly 300 firearms and in excess of 21,000 rounds of assorted ammunition, 567 arrests and the seizure of thousands of pounds of drugs, including cocaine, ganja and hash oil. In addition, some 100 wanted persons have been apprehended, five illegal airstrips disabled and more than 80 illegal aliens detained. For the period January to September [2007], Operation KINGFISH carried out 607 operations and firearms, drugs and ammunition were seized".[4] That same year following the national election, the new Minister of National Security, Derrick Smith, stated "Kingfish has been one of the most successful and celebrated national security projects ever introduced in this country and I wish to commend you and congratulate you wholeheartedly, on behalf of the Government and people of Jamaica."[5] KINGFISH eventually evolved into the Major Organised Crime and Anti-Corruption Agency (MOCA) and still enjoys significant support from the UK.

KINGFISH was a success. Jamaican authorities could have chosen a different path; they could have sent security forces into the drug lord's territories and may well have persevered but undoubtedly there would have been significant violence and loss of life. The population would probably not have supported the government and inevitably innocent Jamaicans would have been caught up in the battles – what are so often euphemistically

---

3 Making a dent in organised crime, *The Gleaner*, 10 November 2005, available at https://gleaner.newspaperarchive.com/kingston-gleaner/2005-11-10/page-8/.
4 *The Jamaica Star Online*, dated 29 October 2007, available at http://jamaica-star.com/thestar/20071029/news/news8.html, searched 3 April 2013.
5 *Ibid.*

referred to as 'collateral damage'. It is not at all clear that such a course of action would have been successful.[6]

The Philippines

On the 10th of November 2013, Super Typhoon Haiyan, a category 5 tropical cyclone and the strongest to ever make landfall, hit the Philippines. The Philippines are comprised of over 7,000 islands and, as one of the most heavily populated countries in the world, are ranked number 117 of 189 nations in the UN's Human Development Index. With wind speeds more than 195 mph, gusts exceeding 235mph, nearly 300mm of rain falling in under 12 hours, 7 metre waves battering the coast and prompting a surge in sea level of nearly 4 metres (13 feet), the Typhoon destroyed everything in its wake.[7] More than 7,000 people were killed, 1.9 million were left homeless and 6 million were displaced. Major rice, corn and sugar producing areas were destroyed and in the immediate aftermath, widespread looting and civil disorder took hold. There would have been worse in less prepared places for the world's biggest typhon to have hit, but not many, and the country buckled under the disaster. As images of destruction and human suffering reached the outside world, the international community began to mobilise.

The UK has a long and honourable tradition of providing international aid, humanitarian assistance and disaster relief through its, then, Department for International Development (DFID), subsequently subsumed in 2020 into the Foreign Commonwealth and Development office (FCDO). In 2010 the UK provided official development assistance (ODA) that amounted to an estimated 0.56 per cent of its gross national income (GNI), the highest amount of ODA to GNI that the UK has provided since the United Nations set a target rate of 0.7 per cent GNI for ODA for the world's wealthiest countries. By 2010 that figure was approaching £8.3 billion (approximately US$13 billion) and for years it has been entrenched in successive governments commitments: "On aid spending our commitment is clear – we won't balance the budget on the back of the world's poorest

---

6 The teams responsible for KINGFISH IO provided the author with a detailed narrative of events but asked to remain anonymous; at some point in the future some of the 'Dons' may be released and even now, eighteen years later, the risk of retribution and revenge cannot be ruled out.
7 Tropical cyclones, BBC bitesize, available at https://www.bbc.co.uk/bitesize/guides/z9whg82/revision/4.

people. Confirming our commitment on aid is both morally right and in our national interest".[8] As a consequence, today the UK is providing foreign aid to over one hundred countries.

International aid is not entirely altruistic and the phrase 'our national interest' is important. The UK, in common with other donors, expects to see a return on its expenditure. What that return is and how it is measured is complex, but we can discern some obvious trends. The UK, in common with many other nations, provides assistance to people in need, in particular, in hostile or conflict states. It does so, DFID states, because "poverty and fragile states created fertile conditions for conflict and the emergence of new security threats including international crime and terrorism."[9] Other reasons for aid include combatting climate change, wealth creation and combating global disease. Again, all of these are obviously benefiting not just the recipients but the UK taxpayer as well, even if the perceived benefit is some way off. With so many deserving causes the UK must prioritise its aid and there is an inevitability that Britain's overseas territories, protectorates and Commonwealth nations will invariantly be looked at favourably.

The Philippines is not a nation in conflict – although it does have a terrorist problem. The Philippines is not a British protectorate, nor is it a member of the Commonwealth and it is not in an area of the world where the UK has an obvious strategic interest. Indeed the Philippines is nearly 7,000 miles from the UK and, with the exception of Hong Kong until 1997, it is in a part of the world in which the UK had seemingly lost interest post the 1956 Suez crisis. The 'big kids on the block' are China, the US and to a lesser extent, Australia. The UK is an infrequent visitor, and the Pacific is a long way from its natural areas of interest in the North Atlantic, The Gulf and the Mediterranean. And so, whilst there should have been no surprise that the UK provided aid assistance, the size and scope of that assistance is surprising. Unless of course one looks at the tragedy not just through the lens of aid but through the looking glass of IO. The truth is that the UK has for some years been seen as a diminishing global power. Its armed forces have shrunk significantly, and its one-time membership of the European Union seen to be more important than perceptions of previous

---

8 House of Commons Library, *supra* note 6, at 15, citing Queen's Speech, International development spending from 2013, Number 10 website (May 2010).
9 *Fighting Poverty to Build a Safer World: A Strategy for Security and Development*, Department for International Development, 2005, available at https://gsdrc.org/document-library/fighting-poverty-to-build-a-safer-world-a-strategy-for-security-and-development/.

'Empire' and influence in the Far East. The US has a sizable naval fleet in the region, controlled by Indo-Pacific Command (INDOPACOM), China is expanding its regional influence significantly through both military expansion and its 'belt and road' initiative and Australia very much sees the region, rightly, as its 'back yard'. And so all were surprised by the scale of the UK's assistance mission to the Philippines – which came to be known as Operation PATWIN. By its end, this comprised an aircraft carrier, a navy destroyer, 11 Royal Air Force aircraft, 16 charted UK Aid aircraft, 25 deployed DFID humanitarian experts, 18 UK medics, the provision of 245,000 emergency shelters, food for over 325,000 people, seeds and tools for over 500,000 people and 100,000kg of rice. Not including the cost of the warships, DFID provided more than £60 million in assistance. Because beyond the altruistic humanitarian shopfront, Operation PATWIN was also a massive Strategic Information Operation to present the UK in a very different way to international audiences and the most important of which was not China but the US and in particular the (then) US Pacific Command (PACOM) in Hawaii. Secondary audiences were Australia, Japan, and thereafter various other regional countries, including China. The message was clear: 'The Brits are back'. To reinforce that, the aircraft carrier HMS Illustrious – the Royal Navy Fleet Flagship – was rendezvoused with the destroyer HMS Daring in the Pacific for a very visible photo opportunity. As ABC News in Australia reported, "This is not just about kindness to an ally."

Indeed, the international reaction to the scale of the UK deployment was initially one of confusion. The [UK] MoD was briefed by various cross government officials that "The initial reaction of [Australian] Defence was incredulous... [Australia] had been too slow to see the strategic communication value".[10] The comments of a senior visiting Singaporean 2* (Major General) military officer were relayed back to the MoD Military Strategic Effects (MSE) department which had largely been behind the information planning for the operation: "Did you think this through in strategic terms or was it just humanitarian altruism? His view was that the assessment by his regional neighbours was that it was the former; actions speak louder than words".[11] They were impressed by the UK's operations we well they should because Operation PATWIN, today, is an exemplar of a good strategic information campaign.

---

10 Comments made during an unclassified post operational debrief in the Ministry of Defence.
11 Ibid.

Colombia

By 2010, Colombia had been embroiled in a bloody insurgency with the Revolutionary Armed Forces of Colombia People's Army (or FARC), a Marxist-Leninist guerrilla group, for nearly 50 years. In 2010, after years of fighting and various unsuccessful attempts to reduce the attraction of the group to new recruits, the Colombian government decided to run an IO that they labelled Operation CHRISTMAS; its intent was to reduce the number of FARC rebels in the jungle. Christmas was chosen because analysis of the intended audiences suggested that FARC rebels – largely Catholic in their up bringing – were particularly vulnerable to psychological pressures. The Colombian military embarked upon one of the most unique and innovative IO in history which would see Special Forces soldiers travel deep into the jungle to key FARC-controlled transit routes and covertly decorate trees with Christmas lights and banners.

Various trip wires and battery packs were installed to support nine large Christmas trees on key routes. At night, when the wires were tripped by FARC rebels picking their way through the jungle trails, the trees suddenly and spectacularly displayed a message in lights: "If Christmas can come to the jungle, you can come home. Demobilize." Alongside the Christmas trees the government engaged in a coordinated media campaign to the families of young FARC rebels, encouraging them to ask their loved ones to leave the jungle for Christmas. The effect on young fighters, far from their homes and families at such an important date in the calendar must have been significant because 331 fighters chose to leave the jungle and demobilise.[12]

The success of the 2010 campaign prompted a second one the following year. Known as Rivers of Light, almost 7,000 glow-in-the-dark floating balls were released into the river, each floating ball filled with thousands of personal messages written by members of river villages, as well as toys, candy, and jewellery in the hopes of encouraging rebels to defect. Once again, the campaign was successful in increasing the demobilisation rate to one person every six hours during the season.[13]

What is not known, at least publicly, is how many of the fighters made their demobilisation permanent, and how many at cessation of the

---

12 Shipley L, *Christmas after Christmas: How a Colombian ad exec helped demobilize guerrillas by advertising peace*, 3 December 2017, available at https://thebogotapost.com/christmas-christmas-colombian-ad-exec-helped-demobilize-guerrillas-advertising-peace/25848/.
13 Ibid.

Christmas holidays, returned to the fight. Regardless, it is an example of an extremely innovative Information Operation.

Ukraine

We have already seen in earlier chapters some examples of Russian IO. But events of 24th October 2014 in central Kyiv in Ukraine, show how cyber-attack and IO can be woven together into very sophisticated operations. The operation, which took place two days before parliamentary elections in Ukraine, saw two digital advertising billboards, one at Raisy Okipnoi Street and the other outside the Lybid Hotel, hacked and aggressive anti-governmental adverts suddenly began appearing. Both billboards are owned by the HitTech digital billboard company based in Kyiv and part of a network of 65 billboards across the country. Both were on busy junctions, controlled by traffic lights. The adverts, each 75 seconds long, accused Ukrainian politicians of war crimes, collateral damage, and the deaths of civilians in the Russian separatist districts of eastern Ukraine.

The imagery was gruesome, dismembered bodies, dead and dying children; all designed not just to shock but to make viewers angry. Angry enough to change their vote. Of note, was that the billboards had been synchronised with the traffic lights – the adverts running as the lights turned red and traffic halted.

Later that day on the Russian VK website, CyberBerkut, an anti-Western hacktivist group which took its name from the Ukrainian riot police and deployed against protesters during the unrest in Kyiv that led to the ousting of President Viktor Yanukovych, released a statement: "Today, we used several dozen billboards in Kyiv to remind the people of Ukraine of the futility of farcical elections… the sooner we deal with the neo-Nazi government and deputies who only profit from this war, the sooner peace and order will be established in the country. We are CyberBerkut! We will not forget! We will not forgive!".[14] The billboard content was then released to mainstream media and on to the internet. Russian-based rusvesna.ru, Russia Today, Rossija 1 and the BBC[15] all covered the story, thus ensuring that the material gained an even wider audience.

---

14 *CyberBerkut hacked Kiev billboards*, VK, 24 October 2014, available at https://vk.com/wall-67432779_14678?lang=en.
15 Shevchenko V, *Ukraine conflict: Hackers take sides in virtual war*, BBC News, available at https://www.bbc.co.uk/news/world-europe-30453069.

The clear intent of the hackers was to dissuade Ukrainians from voting for pro-western politicians. Although it seems that the intended audience was Ukrainians living in Kyiv, the video was also disseminated by small thematic internet portals ensuing that the content received far wider dissemination – and again amplified by mainstream media coverage. It is always difficult to directly attribute outcomes to IO alone, IO is often just one part of often complex issues, but it is perhaps worth noting that the first three parties in that election were all either nationalist or European leaning, collectively accruing over 50 per cent of the vote. The pro-Russian 'Opposition Block' party came forth with just over 9 per cent of votes.

What is probably more relevant is the 'shock' effect the hack would have had on motorists and pedestrians, going about their everyday business and suddenly being unexpectedly confronted by graphic anti-western, pro-Russian digital adverts. If nothing else, it might serve to sow some seeds of concern about the reach that Russia was having into everyday lives inside Ukraine. However, it might also have angered people – perhaps even encouraging some who might not ordinarily have voted.

In May 2018 the UK's National Cyber Security Centre (NCSC) issued a press statement[16] in which it explicitly named CyberBerkut as being part of the Russian Military Intelligence Service, the GRU.

My second example from Ukraine is far simpler. Lilia Aleksandrovna Atroshchenko is the wife of Russian military officer Colonel Sergey Atroshchenko who as well as being the commanding officer of the 960th Aviation Assault Regiment, is now wanted for crimes for ordering the bombing of Mariupol theatre on 16th March 2022 in which up to 700 people died. Lilia likes to send her husband photos of herself either naked or very scantily clad. Unfortunately for her, Ukrainian hackers obtained her email address and a host of other personal details from Russian military websites. That allowed them to contact her by posing as one of her husband's friends in the Regiment, and asking if she could arrange a photoshoot of all the officer's wives as a surprise.

Lilia was only too happy to help. Assured that the photos were exclusively meant for their husbands and not the general public, she arranged a photo shoot of the wives in their husbands best uniform jackets and medals at their Russian military base, together with military backdrops,

---

16 *Reckless campaign of cyber-attacks by Russian military intelligence service exposed*, National Cyber Security Centre, 2018, available at https://cyber-peace.org/wp-content/uploads/2018/10/Reckless-campaign-of-cyber-attacks-by-Russian-military-intelligence-service-exposed-NCSC-Site.pdf.

wide angled shots of the airfield and other useful military intelligence.[17] On March 16th 2023, Lilia sent her photographs from the 'patriotic photo shoot' to her new 'friend' the Ukrainian hackers thereby allowing the positive identification of 12 further soldiers responsible for the Mariupol bombings.[18]

Iraq 2003

The decision to invade Iraq and remove its long-standing President, Saddam Hussein, was hugely contentious and in the UK over 1 million people marched through the city of London demanding that the British government detach British forces from the forthcoming war. Many believed that the justification for war – the so called 'dodgy dossier' presented to parliament and the UN by the government of Tony Blair in February 2003[19] – was itself an IO designed to make the case for war. British Major General, Michael Laurie, told The Chilcott Enquiry that "We knew at the time that the purpose of the dossier was precisely to make a case for war",[20] directly contradicting the then government's claims about the dossier.

We do know that IO were used extensively during the conflict. On 25th March 2003, Brigadier General Vincent Brooks, at a Central Command press briefing, told the media that the US led coalition was broadcasting Radio transmissions, "on five different radio frequencies 24 hours a day and have been doing so since the 17th of February".[21] However, it was not the five frequencies nor the 24-hour transmissions that were new, what was new and what was not mentioned was that many of the extended transmissions originated from coalition ships operating in the Northern Arabian Gulf. The ship born broadcasts – in Arabic and focussed on encouraging Iraqi

---

17 @saintjavelin, Twitter (now 'X'), available at https://twitter.com/saintjavelin/status/1641027966648852480.
18 Hacking a Russian war criminal, commander of 960th Assault Aviation Regiment, InformNapal, 29th March 2023, available at https://informnapalm.org/en/hacking-75387-960-aviation-regiment/.
19 What was to become known as the 'dodgy dossier' began as a proposal of the Iraq Communications Group chaired by Alastair Campbell, who described the purpose of the paper as being "to get our media to cover this issue of the extent to which Saddam Hussein was developing his programme of concealment and intimidation of the United Nations' inspectors. See https://publications.parliament.uk/pa/cm200203/cmselect/cmfaff/813/81308.htm.
20 Norton-Taylor R, *Iraq dossier drawn up to make case for war – intelligence officer*, The Guardian, 12 May 2011, available at https://www.theguardian.com/world/2011/may/12/iraq-dossier-case-for-war.
21 Mäkeläinen M, *Shock and awe on the air. US steps up propaganda war*, DXing, 5 April 2003, available at http://www.dxing.info/profiles/clandestine_information_iraq.dx.

troops to lay down their arms – began at 2300 hours Baghdad time when the purpose built and US PSYOP equipped US EC-130J Hercules Aircraft 'Commando Solo', which had been broadcasting for extended periods, returned to base.

Created by US PSYOPS teams, the broadcast had begun in December 2002 from ships' operating at the mouth of the Khawr Abd Allah waterway, the point where the Shatt al-Arab river emptied into the Persian Gulf. The broadcasts were designed to demoralise Iraqi forces and encourage them to either surrender or simply lay down their weapons and leave the battlefield. As Professor Phil Taylor, a scholar in IO noted, the 21st of March was a significant date. "When the 'shock and awe' bombing of Baghdad began on the night of 20-21 March 2003, Radio Baghdad appeared to have been jammed and words from Commando Solo announced over its frequencies that 'the facilities of the Iraqi regime have started to be hit ...This is the day we have been waiting for ... The attack on Iraq has begun'".[22]

Although the ability to broadcast from ships at sea was not new it was rarely used. But the broadcasts were seen to have been very successful

Iraq 2014

The so-called Islamic State (IS), previously ISIS and briefly ISIL, came to international prominence in the chaos that enveloped Iraq, and subsequently Syria, in 2014. IS's IO were multi-faceted; to encourage the flow of recruits it released multiple videos and messages onto the internet to make the 'Caliphate' an attractive destination for young Muslims and, critically, wealthy Gulf based financiers and supporters. Its third audience group was the hated 'West', and a succession of grizzly videos were released of IS prisoners being executed in the most gruesome fashion. But prior to taking control, IS had used a more tactical IO based around an application called 'Dawn of Glad Tidings'. The app was deceptively simple – once downloaded (from the Google Android Play store[23]) it automatically retweeted any IS tweets it received, inflating and controlling its message. As The Atlantic magazine reported, "Once you sign up, the app will post tweets to your account ... The tweets include links, hashtags, and images, and the same content is also tweeted by the accounts of everyone else who

---

22 Taylor P, *Psychological Operations In Operation Iraqi Freedom, 2003*, available at https://universityofleeds.github.io/philtaylorpapers/vp016d23.html.
23 *Isis official app available to download on Google Play*, itvNews, 17 June 2014, available at https://www.itv.com/news/2014-06-17/isiss-official-app-available-to-download-on-google-play.

has signed up for the app, spaced out to avoid triggering Twitter's spam-detection algorithms".[24]

The app was first used in April 2014, but it was during IS's advance on the city of Mosul that the app really came into its own. As their forces advanced, IS was able to artificially increase the number of tweets generated, generating a 'Twitter (now 'X') storm' of over 40,000 tweets in a day.[25] But this was more than a quest for 'Likes'; the total number of IS fighters approaching the city numbered somewhere between 1,000-1,500 fighters. In contrast, it is estimated that defending the city were more than 20,000 trained and relatively well-equipped Iraqi soldiers – and undoubtedly everyone had their own mobile phone. In the absence of information from their own commanders, the Iraqi troops looked to the internet and social media, discovering to their horror the imminent fall of the city being widely predicted. As a result, many units, in the face of an approaching ISIS army of seemingly far superior numbers, abandoned their equipment and fled. The extent of the Iraqi Army's defeat became quickly clear when officials in Baghdad reported the insurgents had ransacked the main army base of all weapons, had released hundreds of prisoners and "may have seized up to $480 million in banknotes from the city's banks."[26] Worse, two divisions of Iraqi soldiers – over 20,000 men – had turned and run. Local media reported that IS extremists roamed freely through the streets of Mosul, openly surprised at the ease with which they took Iraq's second largest city.[27] Ian Tunnicliffe, a former UK Intelligence Corps Colonel, has studied the use of social media by insurgent and terrorist groups and regards the 2014 IS incident as seminal. Tunnicliffe told me that "the information flow was able to bypass the military hierarchy and be delivered, unfiltered, directly to the soldiers on the ground… It represents one of the earliest documented examples of a successful smartphone and social media enabled tactical/operational psychological operation".[28]

Following the successful attack, IS used social media extensively. For example, it began using platforms such as Ask.fm, a question-and-answer

---

24 Berger J, *How ISIS Games Twitter (now 'X')*, The Atlantic, 16 June 2014, available at https://www.theatlantic.com/international/archive/2014/06/isis-iraq-Twitter-social-media-strategy/372856/.
25 Ibid.
26 Chuvlov M, *Iraq army capitulates to Isis militants in four cities*, 12 June 2014, available at https://www.theguardian.com/world/2014/jun/11/mosul-isis-gunmen-middle-east-states.
27 Ibid.
28 Email Tatham / Tunnicliffe 3 January 2022.

network on which users can post questions and give answers anonymously, to enable supporters to have direct conversations with IS fighters.[29]

Israel 2009

On 27th December 2009, the Israeli Defence Forces (IDF) launched Operation Cast Lead, a mission to end the ability of the Palestinian resistance group Hamas to launch rocket attacks from the Gaza Strip into southern Israeli settlements. Accompanying the ground operations, the IDF also started a concerted information campaign with a strong emphasis on 'New Media' and the very close co-ordination of information across all government agencies. Cast Lead would be the first 'outing' for a new organisation, the Israeli National Information Directorate (NID) that had been established inside the office of the Prime Minister in April 2008. The creation of the NID resulted from failures highlighted in the 2007 Winograd Report,[30] which had criticised the Israeli government for a failure to engage with international audiences during previous military operations.

From the very outset IO were a closely coupled component of the overall military operation. Its objectives were threefold. Firstly, aggressively court positive world opinion and mitigate dissenting voices,[31] secondly, intimidate and/or persuade Palestinian population and finally, reassure the Israeli population. Key elements of Israel's strategy were to have close control of domestic and foreign journalists within Gaza, a proactive focus upon online and new media engagement and the deployment of fluent Arabic and English speakers as global spokespeople. Initially Israel was confident that it had gained the upper hand.

In the initial information shaping phase it would seem that Israeli objectives were met. The French newspaper Le Figaro undertook a survey and revealed that 55 per cent of respondents were "understanding towards the Israeli operation" whilst former UN Ambassador, Dan Gillerman, suggested that "Israel has no small measure of understanding and

---

29 Marks P, *How ISIS is winning the online war for Iraq*, The New Scientist, 25 June 2014, available at https://www.newscientist.com/article/dn25788-how-isis-is-winning-the-online-war-for-iraq/.
30 The commission of inquiry into the events of Israeli conflict in Lebanon 2006 chaired by retired Israeli judge Eliyahu Winograd.
31 This objective was publicly stated by an IDF Officer to NATO School Oberammergau during the NATO Senior Officers StratCom course, September 2013.

support".[32] However, once the actual ground offensive started on 3rd January 2009, it appeared to be able to do little about the damaging claims that hundreds of civilians, including children, had been killed by IDF forces. Images of dead Palestinian children, reinforced by IDF strikes on a UN school and later claims that the IDF were committing war crimes by using white phosphorus against civilian targets, created significant global public protest against Israel in the second and third weeks of the offensive.

The violence and virulence of these anti-Israel and frequently anti-Semitic protests that followed, as well as the deep sympathy for Hamas, shocked Israeli public opinion.[33] Dominique Wolton, a media specialist at France's National Centre for Scientific Research (CNRS) in Paris noted that: "Israel are the ones who have a grip on communications, but Israel will not win the communications battle because, whatever Israel's legitimate rights are, the unbalanced use of force and the unleashing of violence by Israel is acting against it". Charles Tripp, Professor of Middle East Politics at the School of Oriental and African Studies (SOAS) in London took a similar view when he stated that: "the very powerful images of what's happening to civilians in Gaza must be having a greater impact than seeing Israeli spokesmen talking about the war on terror. In many ways, one of the main targets of the Israeli propaganda is Europe and the US, and I would have thought they're not doing too well there". [34]

However, Israel's decision to recognise the importance of information, and resource it properly, made wartime messaging much more cohesive, far quicker to react to allegations and more focused. The fact that the IDF chose to conduct at least part of their operation in the information environment – and to invest in new technology and new conduits, to engage new, perhaps younger, audiences without any guarantee of success – is significant. Crucially they did so in a co-ordinated manner across the government. It is also worth considering how the Israeli position would have looked to the international community if they had not engaged in an information campaign.

---

32 Shabi, Rachel. *Special Spin Body gets Media on Message*. The Guardian. 2 January 2009, available at http://www.theguardian.com/world/2009/jan/02/israel-palestine-pr-spin accessed 11 December 2014.
33 Sokol Sam. *Coming to Defense on all Fronts*. 29 July 2014. The Jerusalem Post, available at http://www.jpost.com/Jewish-World/Jewish-Features/Jewish-communities-begin-to-quantify-wartime-anti-Semitism-upswing-369300.
34 *Israel Propaganda takes International Beating*. Middle East Online 7 January 2009, available at http://www.middle-east-online.com/english/?id=29623.

Gaza 2023/2024

At 0630 am on 7th October 2023, Israeli's woke up to sirens warning of imminent attack. An estimated 2,200 missiles were subsequently fired into southern and central Israel. "On motorbikes, by car and on foot, bristling with weapons, a first wave of 400 Hamas militants poured across the border into Israel at the 15 points where they had breached the security barrier".[35] Attacking settlements along the southern border and killing an estimated 1200 people, the Hamas fighters subsequently abducted a large number of men, women and children, taking them back to Gaza as hostages. Shortly afterwards Mohammed Deif, commander in chief of Hamas' military arm, Al Qassam Brigades, released a video statement claiming responsibility for the attack.[36] In response the Israeli government, stung by the clear failure in intelligence, officially declared war, the first time the state has done so since 1973. In Operation Swords of Iron, Israeli Defence Forces (IDF) launched sustained attacks across the Gaza Strip, designed to destroy Hamas's military and governing capabilities and free the hostages.[37]

The physical battle taking place, in the tunnels and towns of Gaza, was only one part of the wider conflict; predictably both sides quickly began their IO. So too other actors. On 9th October, just two days after Hamas' attack, stories began circulating in the Ukrainian media that Russia had given Hamas EU and US manufactured weapons captured from Ukraine forces.[38] The newspaper reported that: "defence Intelligence believes that the Russian occupiers intend this fake news to form the basis of a series of 'revelations' and 'investigations' in the Western media... the aim of the Russian provocation is to discredit the Armed Forces of Ukraine and bring the flow of military aid to Ukraine from Western partners to a complete stop".[39] As ever the truth is elusive but had it been true (and the

---

35 Beaumont P, *How did Hamas manage to carry out its rampage through southern Israel?*, The Guardian, 9 October 2023, available at https://www.theguardian.com/world/2023/oct/09/how-did-hamas-manage-to-carry-out-its-rampage-through-southern-israel.
36 Hutchinson B, *Israel-Hamas War: Timeline and key developments*, ABC News, 22 Nov 23, available at https://abcnews.go.com/International/timeline-surprise-rocket-attack-hamas-israel/story?id=103816006.
37 *What is Hamas and why is it fighting with Israel in Gaza*, BBC News website, available at https://www.bbc.co.uk/news/world-middle-east-67039975.
38 Mazurenko A, Ukrainska Pravda, *Russians gave Hamas trophy weapons to discredit Ukraine – Defence Intelligence*, 9 October 2023, available at https://www.pravda.com.ua/eng/news/2023/10/9/7423273/.
39 Ibid.

logical presumption is it was not), it would have destroyed US and wider international support for Ukraine.

A few days after this event, evidence of the Israeli IO plan began to emerge and a colleague in the region sent me what looked to be part of an Israeli government and/or IDF directive on messaging, audiences and themes. Although difficult to verify, my contact's credentials were impeccable and, besides, it looks exactly like the type of Strategic Communication planning and directive that I would expect to see in this type of conflict.[40]

In the summary paragraph the document stated that the objective was to 'create empathy for the Israeli people in global public opinion'. It identified the target audience as non-Jews in western democracies and stated that the method would 'control of the narrative by emphasis personal stories accompanied by touching videos and photos'. The document then provide some 'dos' and 'donts' of which the most notable were 'don't draw the discourse into a discussion about the Israel-Palestinian conflict as a whole, rather focus on hostages, the elderly, children' and 'do document our day to day life – the people standing in line to donate blood, aid packages being packed for soldiers...'. Two incidents early in the campaign exemplify that information battle and put the Israeli guidance to significant test: the alleged decapitation of babies and the alleged attack on the Al-Ahli hospital.

The decapitated baby story emerged from a claim, allegedly made during an IDF press facility on 10th October, that Hamas had beheaded babies during their attack on the Kfar Aza Kibbutz. The claim, which appeared to originate with the Tel Aviv-based news channel i24,[41] was that 40 babies had been decapitated. Despite other journalists on the press facility being unable to confirm the allegation,[42] it spread rapidly and gained widespread global attention on social media, amassing over 40 million impressions on Twitter (now 'X') alone. Despite the IDF being unable to immediately

---

40 I posted this document to my LinkedIn profile where it can be viewed: https://www.linkedin.com/feed/update/urn:li:activity:7118927618455805952/.
41 Sky News, *What we actually know about the viral report of beheaded babies in Israel*, 12 October 2023, available at https://news.sky.com/story/its-important-to-separate-the-facts-from-speculation-what-we-actually-know-about-the-viral-report-of-beheaded-babies-in-israel-12982329.
42 For example, see Le Monde's Jerusalem based reporter @samforey Twitter 11 October 2023 11:18 or Israeli phot journalist Oren Ziv who tweeted that 'during the tour we didn't see any evidence of this, and the army spokesperson or commanders also didn't mention any such incidents' @orenZiv_ 12:30 11 Oct 23.

substantiate the claim,[43] the Israeli PM's official spokesman confirmed to the media that babies had indeed been found with their heads decapitated. US President Biden gave the story further credibility when he told a press conference he had been shown pictures of decapitated children. However, later that same afternoon, the Whitehouse issued a retraction saying that the President had in fact not seen pictures of decapitated children[44] and on 12th October CNN reported that Israel was unable to confirm the specific claim, contradicting their official spokesman.[45]

Other horrific stories followed. One of the search and rescue personnel first on the scene, Yossi Landau, told of his horror at finding a pregnant woman with her stomach cut open and the foetus exposed – an allegation that the UK's Daily Express newspaper headlined as 'Hamas sliced baby out of pregnant Israeli's womb'. A deeply distressing video was circulated online that purported to show a pregnant woman being disembowelled however even Israeli newspaper Haaretz was suspicious, stating that the story was false and "police have no evidence of a body".[46] Other stories that emerged included that Hamas had roasted babies alive in ovens and many women claimed they had been raped.

It is impossible to determine the veracity of all of these allegations. What is unequivocally clear is that Hamas fighters killed a large number of non-combatants in a hideous and brutal act of terrorism. But did they behead 40 babies? The balance of the evidence appears to suggest that they did not. However, one of the immediate consequences of these various allegations was that the official lexicon describing Hamas changed almost over-night with the insertion of the word 'ISIS' into any official statement. Indeed very quickly #HamasisISIS began to trend on social media. This was a very helpful term for the Israeli government. The Palestinian conflict is to many in Europe and the US utterly impenetrable but collective

---

43 Speri A, *Beheaded babies" report spread wide and fast — but Israel military won't confirm it*, The intercept, 11 Oct2023 available online at https://theintercept.com/2023/10/11/israel-hamas-disinformation/.

44 Nelson & Nava, *Biden did not actually see 'confirmed pictures of terrorists beheading children' as he claimed, WH clarifies*, The New York Post, 11 Oct 2023 available at: https://nypost.com/2023/10/11/biden-ive-seen-pictures-of-terrorists-beheading-children-in-israel/.

45 See @cnni Twitter account Oct 12 2023 2:21pm and @sarasidnerCNN Twitter account 12 Oct 23 1.29pm.

46 Hasson, Rozovsky, Haaratz, *The Hamas massacre led to the spread of horror stories, not all of which happened in reality. The truth is hard enough*, 3 December 2023 (in Hebrew) available online at https://www.haaretz.co.il/news/politics/2023-12-03/ty-article-magazine/.premium/0000018c-2036-d21c-abae-76be08fe0000.

knowledge of the brutality of ISIS is far more recent. The images of the so called 'Beatles', British jihadis fighting for ISIS, cutting off the heads of western aid workers on video, was widely reported in western media and in associating Hamas with a known and feared terrorist entity such as ISIS, the Israeli government achieved a notable success in shaping the perception of western audiences. Arguably however that advantage was lost as the consequences of the Israeli military campaign in Gaza began to slowly become apparent.

Just ten days later another event occurred that would again dominate the international media – the explosion at the Al-Ahli Arab Hospital. The Gaza Health Ministry reported 342 injured and 471 killed whilst the Anglican diocese that manages the hospital reported 200 people killed.[47] Very quickly Hamas blamed the Israelis. Israel, meanwhile, quickly released a video via its @Israel Twitter (now 'X') account, purporting to prove the strike was a result of a malfunctioning Palestinian rocket, only for a former Bellingcat analyst and now *New York Times* reporter @AricToler to discredit it, who demonstrated from the video's time code that it had been recorded 40 minutes after the hospital strike. The video was removed by the @Israel account immediately. Egyptian President Abdul Fattah Al-Sisi denounced in the strongest terms the "intentional bombing" of the Gaza hospital, further describing it as "a flagrant violation of international law".[48] The United Arab Emirates and Bahrain, both of whom who had normalised ties with Israel in 2020, were quick to condemn the 'Israeli attack' on the hospital.[49] Even Oman, one of the most moderate of all Arab states, condemned "the targeting of the hospital by the Israeli occupation… a war crime, genocide and violation of the international and humanitarian law".[50] Finally, regional hegemon Saudi Arabia denounced the "heinous crime committed by Israeli occupation forces".[51]

Like the rapes, decapitations and burnt babies it seems an almost impossible task to definitively conclude what happened but in their

---

47 https://www.hrw.org/news/2023/11/26/gaza-findings-october-17-al-ahli-hospital-explosion.
48 @AlsisiOfficial Twitter 17 Oct 2023 9:55pm in Arabic.
49 @UAEMissiontoUN Twitter 18 Oct 2023 2:54am.
50 The New Arab, *Arab countries blame Israel for Al-Ahli Hospital massacre amid anti-US backlash*, 18 October 2023, available at https://www.newarab.com/news/arab-countries-blame-israel-al-ahli-hospital-massacre.
51 Saudi Press Agency, *Foreign Ministry: Saudi Arabia condemns in strongest terms Israeli Occupation Forces' Heinous shelling of Al-Ahil Baptist hospital in Gaza*, 17 October 2023, available at: https://www.spa.gov.sa/en/N1981747.

investigation into the incident, Human Rights Watch found that the "possibility of a large air-dropped bomb, such as those Israel has used extensively in Gaza, highly unlikely" and noted that "the rockets that are extensively used by Palestinian armed groups such as Hamas and Islamic Jihad have only rudimentary guidance systems and are prone to misfire, making them extremely inaccurate and thus inherently indiscriminate when directed toward areas with civilians".[52] In a sense though the truth is irrelevant. Huge swathes of the 'Arab street' will believe that Israel bombed the hospital deliberately, regardless of evidence, and huge swathes of the Israeli and Jewish diaspora will believe that Hamas decapitated babies, regardless of the evidence base.

---

52 Human Rights Watch Report. *Gaza: Findings on October 17 al-Ahli Hospital Explosion 26* November 2023 *"Evidence Points to Misfired Rocket but Full Investigation Needed"*, available at https://www.hrw.org/news/2023/11/26/gaza-findings-october-17-al-ahli-hospital-explosion.

7 Russia and Ukraine

In 2017 Marat Mindiyarov, a 44-year-old Russian with teeth only a dentist could love, gave an interview to the US' PBS News Hour investigation, *Inside Putin's Russia*, outside the offices of the so-called Internet Research Agency (IRA) at 55 Savushkina Street, St Petersburg, Russia. Owned and run by the late Russian oligarch Yevgeny Prigozhin (of Wagner Group infamy), the IRA's goal is to influence online engagement in favour of Russian political and economic interests. Thought to have been started in 2013 to stifle internal dissent, the agency soon expanded its operations. Mindiyarov told PBS that each day he and his colleagues would receive documents directing the day's work schedule. On 24th December 2014 he was told to "create a negative attitude about [President] Obama's foreign policy". That day he created and posted photographs comparing Obama to Hitler, cartoons of the US as a giant fish about to eat up the planet and a cartoon of the American eagle sharpening its talons with a file. He then posted a question: "Can the US take Russia Out" on 50 different websites in 23 cities across Russia. Over the next 12 hours his posts were picked up and reposted over 600 times from 70 fake accounts. Mindiyarov told PBS that he had perhaps 20-30 different online accounts and that some colleagues of his had 'maybe in the hundreds', all tweeting and posting to directed story lines. "Everything is very simple there. Black and White. No colours. Just black and white".

Mindiyarov is not the only former employee to break cover. In 2015 a fellow IRA employee Lyudmila Savchuk told the UK's Sunday Telegraph newspaper that her job was to spend 12 hours a day praising the Kremlin and lambasting its enemies on social networks, blogs and in the comments section of online media. "We had to say that Putin was a fine fellow, that Russia's opponents were bad and [the monkey] Obama was an idiot". Provided she met her target of 135 pro-government comments per shift she earned 45,000 roubles a month (equivalent to £450).

The IRA's efforts appear to extend far further than just a few social media posts. Take, for example, the 11th of September 2014 announcement (on Twitter, now 'X') of a powerful explosion occurring at a chemical manufacturing plant in Centreville, Louisiana in the US.[1] This was quickly followed by text messages to local residents saying that toxic gas had been released from the plant as a result of the explosion. New York Times Magazine journalist, Adrian Chen, describes what happened: "a highly coordinated disinformation campaign, involving dozens of fake accounts that posted hundreds of tweets for hours, targeting a list of figures precisely chosen to generate maximum attention. The perpetrators didn't just doctor screenshots from CNN; they also created fully functional clones of the Web sites of Louisiana TV stations and newspapers ... an effort of this scale must have taken a team of programmers and content producers to pull off".[2]

In an effort to stem rising public concern, the chemical company issued a statement. Timothy Fedrigon, the Deputy Chief People Officer for the Birla Carbon company, told the media that: "there has been no release of such toxic gas, explosion, or any other incident in our facility. We are not aware of the origin of this text message. Law enforcement authorities have been contacted and are following up on this matter."[3] The IRA's work?

Pandemics and infectious diseases have proven a rich historical theme of Russian disinformation. On 17th July 1983, The Patriot, a small circulation, Indian newspaper that, according to former KGB officer and defector Ilya Dzirkvelov,[4] had been set up by the KGB in 1962 "to publish disinformation", printed an anonymous letter headlined "AIDS may invade India: Mystery disease caused by US experiments." Allegedly written by a 'well-known American scientist and anthropologist' the article stated that "AIDS...is believed to be the result of the Pentagon's experiments to develop new and dangerous biological weapons."[5] The article cited Fort Detrick, a US Army Medical installation in Frederick, Maryland, which between 1943 and 1969 was the centre of the US biological weapons program, as the origin of AIDS. It claimed that the disease was manufactured from a number of different

---

[1] Crovitz L G, *Putin Trolls the U.S. Internet*, The Wall Street Journal, 7 June 2015, available at https://www.wsj.com/articles/putin-trolls-the-u-s-internet-1433715770.
[2] Chen A, *The Agency*, New York Times Magazine, available at https://www.nytimes.com/2015/06/07/magazine/the-agency.html.
[3] *Company issues statement following text messaging hoax*, KATC.Com, 12 September 2014, available at https://web.archive.org/web/20140912000559/http://www.katc.com/news/company-issues-statement-following-text-messaging-hoax/.
[4] US State Department, *Soviet Influence Activities: A Report on Active Measures and Disinformation, 1986-1987* available to download at https://jmw.typepad.com/files/state-department---a-report-on-active-measures-and-propaganda.pdf.
[5] Boghardt T, *Soviet Bloc Intelligence and its AIDS disinformation Campaign*, Studies in Intelligence, Vol 53, No. 4 (December 2009).

viruses collected by the US in Africa and Latin America. Various versions of this story appeared across the US in the following years. In 1984, for example, US dermatologist Alan Cantwell claimed that the AIDS virus was man-made and had originated from Hepatitis B vaccine trials in New York City; "gays were used as guinea pigs for trials...I don't believe American AIDS came from Africa as they would have you believe, it traces back to the same year, same geographic site, same population that was exposed to government experiments. Young healthy gay men in New York City".[6]

In August 2014, Yahoo News' Twitter (now 'X') account was hacked, and a tweet issued which stated, "Ebola Outbreak in Atlanta. Estimated 145 people infected so far since Doctors carrying the disease were flowing in from Africa".[7] Yahoo moved quickly and within ten minutes had deleted the tweet, issuing a statement that the earlier tweet was unauthorised and contained misinformation. How could something like that happen? Prank? Dissatisfied employee? Or something more sinister? The fact that it involved a deadly, communicable disease points to the latter. Indeed the Ebola theme has been recurrent; rumours that the disease had been brought to the United States have been particularly prevalent in right-wing conspiracy sites, with websites such as 'Gateway Pundit' and 'Breitbart' both claiming that the Ebola virus was confirmed in Texas, brought in by a surge of illegal migrants, whilst controversial Fox TV News host, Laura Ingraham, argued on air that there is a danger of people with Ebola being allowed into the United States under its 'weak asylum laws'. Sylvie Briand, the director of infectious hazard management at the WHO's Pandemic and Epidemic Diseases Department called the challenge of false rumours an 'infodemic'.[8]

In 1986, German scientist, Jakob Segal, concluded that AIDS had been engineered by the US government and tested on prison inmates as early as 1978 and part of a plot by the US government to deliberately target African Americans[9] – a statement subsequently described as 'nonsense' by other German AIDS experts.

---
6 Dr. Alan Cantwell – *AIDS and the Doctors of Death – Part 1*, Yippee YouTube channel, available at https://www.youtube.com/watch?v=6-BlE9dWqWU.
7 Smith S, *No Ebola Outbreak in Atlanta, Twitter hacked*, iMediaEthics, 10 August 2014 available at https://www.imediaethics.org/no-ebola-outbreak-in-atlanta-yahoo-news-twitter-hacked/.
8 Wilson R, *Ebola outbreak in Africa spreads fake news in America*, The Hill, 12 June 2019, available at https://thehill.com/policy/international/448197-ebola-outbreak-in-africa-spreads-fake-news-in-america.
9 Heller J, *Rumours and Realities: Making Sense of HIV/AIDS conspiracy narratives and contemporary legends*, American Journal of public health, January 2015 105(1) e43-e50.

In 1992, then-Director of the Soviet Union's Foreign Intelligence Service (SVR), Yevgeni Primakov, intimated that the KGB was behind the campaign,[10] falsely claiming that AIDS was created by the US government, part of a Soviet attempt to foster distrust and hatred globally of the US. Later, Soviet newspapers publicized the false AIDS-made-in-US claims of a Soviet citizen living in East Berlin, retired biophysicist Dr Jakub Segal. Segal may have been chosen by the East Germans because not only was he the former head of East Germany's Institute for Applied Bacteriology – and therefore a 'credible' voice – but he was a staunch supporter of the then East German regime, an established KGB contact and a known informer.[11] Whether Cantwell had a connection to the USSR, or was acting on genuinely believed principles, is unknown but collectively his and the many other 'authoritative' voices (so called 'useful idiots') helped perpetuate a belief that even as late as 2015 researchers found it had lingered within certain sections of the U.S. African American community that still believed: "HIV/AIDS was a genocidal government plot against African Americans".[12]

It is perhaps odd that what goes on within 55 Savushkina Street is now so well known; indeed to those in 'the know', its public profile, and the ease with which foreign journalists can visit and film, is not at all in keeping with Russia's tradition for secrecy. Which has led many to believe that whilst it may have been a functioning disinformation centre, and it may still be so (although it is far from clear that this is the case with some press coverage in 2020 suggesting that it has either relocated or outsourced its operations to Ghana and Nigeria[13]), it is almost certainly a helpful distraction from the real business of Information Warfare. Russia's Information Warfare is being run not just by quasi-private bodies, of which there are probably two or three more inside Russia (and the town of Obninsk is occasionally cited as a possible location for at least one such centre) but more importantly by the Russian armed forces themselves.

The Russian Army has undergone huge changes since the demise of the Soviet Union in 1991. The US Defence Intelligence Agency assessed

---

10 According to former US State Department Official, Todd Leventhal who had responsibility for countering Russian disinformation, Primakov did not admit this specifically, only that the KGB had planted false stories blaming AIDS on U.S. military experiments. However believing they planted the Patriot article was considered by the US to be a reasonable extrapolation from what he had said. Source Email Tatham / Retired US State Department official 10 February 2024.
11 Boghartd T, ibid.
12 Heller J, ibid.
13 Ward C et al, *Russian election meddling is back -- via Ghana and Nigeria -- and in your feeds*, CNN, 11 April 2020, available at https://edition.cnn.com/2020/03/12/world/russia-ghana-troll-farms-2020-ward/index.html.

Russian defence spending at its peak in 2016 as $61 billion (around 4.5 per cent of GDP). By 2020 that investment had allowed Russia to build an armed forces of just over 1 million men and women, the world's fifth largest after China, India, the US, and North Korea. The official Russian news agency, TASS, noted in 2018 that: "civilian staff, including... psychologists and doctors, constitutes about 40% of the Armed Forces personnel".[14] Size alone is not necessarily an indicator of quality (as the 2022 Ukraine War more than amply demonstrated) and is widely accepted that whilst elements of the military are very capable – notably their Special Forces – they also have significant weaknesses, particularly in modern military technology such as drone design and production, electronic warfare and radar and satellite reconnaissance. Much of this is a direct result of the post-Crimea embargo imposed upon Russia which saw France and German assistance, in particular, cease. It is also of note that even today conscripts make up over one third of the military total numbers with all the attendant problems and challenges that a conscript service model brings. But what of IO?

Russia does not use the term IO, instead its military doctrine speaks of Information Confrontation and the Information Battleground[15] where [different sides] "seek to inflict defeat (damage) on the enemy through information influence on its information sphere (including information, information infrastructure, entities engaged in the collection, formation, dissemination and use of information, as well as the systems that govern the resulting social relations), while countering or reducing such an impact on its part".

We know that in summer 2013, at a meeting of the Russian Security Council, President Putin mentioned that "information attacks are already being used to solve military-political problems. Moreover, according to experts, their so-called striking power can be even higher than the one of the conventional weapons."[16] On the same day 'a source' in the Russian Ministry of Defence told RIA Novosti (the state-owned news agency) that by the end of 2013 the Ministry: "planned the creation of a separate military branch responsible for the information security of Russia. The main tasks of these new troops will be monitoring and processing of information coming from abroad, as well as fight against the cyber threats. Their officers will

---

14 *Size of Russian armed forces decreases by nearly 300 personnel*, 1 January 2018, available at https://tass.com/defense/983867.
15 ИНФОРМАЦИОННОЕ ПРОТИВОБОРСТВО available at https://encyclopedia.mil.ru/encyclopedia/dictionary/details.htm?id=5221@morfDictionary.
16 Available in Russian at http://special.kremlin.ru/events/president/transcripts.

be obliged to learn a foreign language, first of all English".[17] In May 2014 TASS reported that, according to another source in the Ministry of Defence, "the troops [sic] of IO" were created in the Russian armed forces. Their goal is information and cyber combat, and defence of the military information systems from cyberattacks. The article went on to say that whilst plans had been in the making for some years, the revelations by Edward Snowden of the US' global electronic surveillance systems had accelerated the decision making. Snowden had been a CIA employee, but his revelations were made when he was a contractor for Dell and then Booz Allen, assigned to work with the National Security Agency. The article closed with the statement that "The head of the new structure has been appointed, a military leader with a general's rank".[18] One unit that has come to public prominence is Russian military unit 54777, also seemingly known as the 72nd Special Service Center[19] and the Foreign Information and Communication Service of the Main Directorate of the General Staff. 54777 is a military intelligence psychological operations unit that has been particularly proactive in its efforts in the West, particularly since the Russian invasion of Ukraine.[20] 54777 is believed to have been particularly active during the invasion of Crimea in 2014, providing 'plausible deniability' and 'shaping the narrative'.[21]

Where they sit in Russia's military command and control is unclear, although many Russian analysts believe them to be nested into the newly established directorate for political-military affairs. And who commands them is again unconfirmed but various papers and analysts have linked them with the Deputy Defence Minister General, Andrey Kartapolov, who had headed up Russian military operations in Syria and was the former commander of the strategically important eastern military district that abuts NATO and the west.[22]

---

17 РИА Новости, *Минобороны может создать отдельный род войск по борьбе с киберугрозами* 05.07.2013 , https://ria.ru/20130705/947802340.html.
18 ТАСС, *Источник в Минобороны: в Вооруженных силах РФ созданы войска информационных операций* 12 МАЯ 2014, https://tass.ru/politika/1179830.
19 *Russian military intelligence behind spread of disinformation*, Parliamentary Question European Parliament, 18 June 2020, available at https://www.europarl.europa.eu/doceo/document/E-9-2020-003669_EN.html.
20 *Russia's military unit 54777, disinformation and psychological operations abroad*, Insight Media, 11 January 2023, available at https://insightnews.media/russias-military-unit-54777-disinformation-and-psychological-operations-abroad/.
21 Soldatov & Weiss, *Inside Russia's Secret Propaganda Unit*, 7 December 2020, available at https://newlinesmag.com/reportage/inside-russias-secret-propaganda-unit/.
22 Treyger E, Cheravitch J, Cohen R, *Russian Disinformation Efforts on Social Media*, RAND, available at https://www.rand.org/content/dam/rand/pubs/research_reports/RR4300/RR4373z2/RAND_RR4373z2.pdf.

What is perhaps clearer is what they exist to do – "Information confrontation is an integral part of relations and a form of the struggle between sides (each of which seeks to inflict defeat (damage) on the enemy through information influence on its information sphere, while countering or reducing such an impact on its part.... informational-psychological impact on the population and individuals".[23] As British analyst, Kier Giles, notes in the NATO handbook of Russian Information Warfare: "In the Russian construct, information warfare is not an activity limited to wartime. It is not even limited to the "initial phase of conflict" before hostilities begin, which includes information preparation of the battle space. Instead, it is an ongoing activity regardless of the state of relations with the opponent; in contrast to other forms and methods of opposition, information confrontation is waged constantly in peacetime". Or as Putin advisor Vladislav Surkov (First Deputy Chief of the Russian Presidential Administration from 1999 to 2011), said with regard to the post-2014 war in Ukraine: "The underlying aim is not to win the war, but to use the conflict to create a constant state of destabilized perception in order to manage and control". Chief of the Russian General Staff, Valerii Gerasimov, wrote in 2013 that "The role of non-military means of achieving political and strategic goals has grown, and, in many cases, they have exceeded the power of weapons in their effectiveness".[24]

Why does Russia place such importance upon IO? Like the West, it recognises that the information environment is now a contested space, but the Kremlin also looks back at a military history that is far from glorious and sees information as a 'force multiplier' to ensure that history is not repeated. As is so often the case in modern Russian history one must start with the 1917 winter revolution to find the antecedents of today's IO troops. It was Vladimir Lenin himself who allegedly declared that to explain the revolution's ideas and principles to a largely uneducated (and in many instances illiterate) population, "We can and must write in a language which sows amongst the masses hate, revulsion and scorn towards those that disagree with us".[25,26] At the height of World War II, in which the Soviet Union incurred more losses than any other nation, a 1943 Communist

---

23 информационное противоборство' Russian MoD's official website, available at https://encyclopedia.mil.ru/encyclopedia/dictionary/details.htm?id=5221@morfDictionary.
24 Gerasimov, V. (2013) 'Новые вызовы требуют переосмысления форм и способов ведения боевых действий,' Военно-промышленные кур'ер, No. 8.
25 O'Meara A, *Opening the American Mind: Recognising the Threat to the Nation*, Xlibris, 2016.
26 Lindley-French J, *Vladimir Putin and the October Revolution*, 24 October 2017 available at https://lindleyfrench.blogspot.com/2017/10/vladimir-putin-and-october-revolution.html.

party directive read: "Members and front organisations must continually embarrass, discredit and degrade our critics... constantly associate those who oppose us with those names which already have a bad smell.[27] As the Cold War settled over Europe a growing realization dawned in the Kremlin that Russia could not compete with the US in technology and research and development, and so attention was paid to finding alternative ways to secure influence and exert power.

Russia cultivated two complimentary theories of warfare: 'Reflexive Control'[28] (the art of making your opponent do what you want them to) and 'Maskirovka'[29] (the art of deception and confusion). Actively seeking alternatives to Hard Power and using newly developed computer technology, Soviet scientists were directed to consider how the USSR might interfere in the psychology of political and military decision making. One of its architects was a mathematical psychologist, Vladimir Lefebvre, and he described his work as a process to: "influence his [the adversary] channels of information and send messages that shift the flow of information in a way favourable to us."[30] In today's globalized information environment, where social media is so prevalent and [mis]perception can quickly becoming people's reality, Lefebvre's 1968 quote looks remarkably prescient.

Russia learnt a great deal about Information Confrontation during its ultimately ill-fated mission to Afghanistan. In the Russian journal of Military thought, Colonel Yu. Ye. Serookiy reflected on what Afghanistan had taught the Russian military: "Psychological warfare has become a full-fledged component of modern operations plans which in some cases decides the subsequent course of combat action".[31] The first Gulf War in 1991 was a further wakeup call to the Kremlin. Iraq's air defence network – known as Kari – was patterned on soviet technology and thinking.[32]

---

27 https://www.reddit.com/r/JordanPeterson/comments/hymv0f/for_some_reason_this_sounds_very_familiar/.
28 Reflexive control is defined as a means of conveying to a partner or an opponent specially prepared information to incline him to voluntarily make the predetermined decision desired by the initiator of the action.
29 It is a term that has no direct equivalent in the West but it simultaneously encompasses the arts of: "concealment, the use of dummies and decoys, disinformation and even the execution of complex demonstration manoeuvres. Indeed anything capable of ... weakening the enemy".
30 Daily & Parker (Ed) *Soviet Strategic Deception* (Hoover Press, 1987), 294.
31 Serookiy, Psychological-information warfare: lessons of Afghanistan Military Thought 2004-03-31MTH-No. 001 Pages:196-200.
32 Grahma B, *Gulf War Left Iraq Air Defence beaten not bowed*, The Washington Post, 6 Sept 1996, available at https://www.washingtonpost.com/archive/politics/1996/09/06/gulf-war-left-iraqi-air-defense-beaten-not-bowed/97b33488-b07f-4a32-834f-076fa5bd0f70/.

It was self-evidently no match for the US led coalition and was quickly overwhelmed. And, like the West, the role of the media in what was the first 'real time' TV war, with images of the conflict broadcast across the world as they happened, caused a profound reassessment of military thinking and doctrine. This was quickly followed by the first Chechen war in 1994 which proved deeply uncomfortable for the Kremlin; their Counterinsurgency (COIN) campaign largely failed, and they were alarmed at the success the Chechen separatist narrative seemed to have in the international community. Russia's strategy appeared to transform the crisis from a small nationalist rebellion into a widespread jihadi insurgency which encouraged ever more terrorism and violence. As commentators observed: "the Russian counterinsurgency has been unsuccessful, as the insurgents are neither demolished as a force nor are they isolated by society. Losing the hearts and minds among the Chechen people is a key reason behind why the Russian operation in Chechnya suffered failures. Too little attention was paid to winning over the "hearts and minds" of the people."[33] The Kremlin ended the twentieth-century watching its key ally, Serbia, become a pariah within the international community.

The next two events were to have a profound effect on the Kremlin: The Global War on Terror (GWOT) began with the September 11 attacks on the World Trade Centre in 2001 and the subsequent Coalition deployment to Afghanistan and the 2003 Iraq invasion that followed, and then the operations against Georgian President Saakashvili in 2008.

The New York City attack horrified Moscow. President Putin was one of the first foreign leaders to speak directly to President Bush; he expressed his condolences and offered his unequivocal support for whatever the US might chose to do next.[34] One of Putin's harshest internal critics, Boris Nemtsov, leader of the Union of Right Forces in the State Duma, told the media that he had changed his position from a 'dove' to a 'hawk' over the war in Chechnya. Nemtsov's daughter, Zhanna, had been in New York City on the day of the attack: "there is no sense in even pronouncing the word 'negotiations.' All conversations should be conducted only in the language

---

33 *Younkyoo Kim, Blank S, Insurgency and Counterinsurgency in Russia: Contending Paradigms and Current Perspectives,* Studies in Conflict & Terrorism, 36:11, 917-932.
34 McFaul M, *US-Russian Relations after September 11 2001,* Carnegie Endowment for International Peace, available at https://carnegieendowment.org/2001/10/24/u.s.-russia-relations-after-september-11-2001-pub-840.

of a 'Kalashnikov.'"³⁵ In the various wars that followed – notably Iraq and Afghanistan – the Kremlin watched carefully how the US and its allies struggled to adjust to an adaptable enemy that used information as one of its major weapons. At the same time, in Georgia, the Kremlin scored a notable success.

The UK MoD's Advanced Research and Assessment Group (ARAG) believed Reflexive Control techniques were extensively used to destabilize the Georgian President. Long time Russian analyst, Lt Col Charles Blandy, believed that Saakashvilli's character and personality made him ripe for a reflexive control operation. He was: "hot headed, often rash in his decisions and intemperate. In my view the Russians knew how to raise the political pressure on him and also what he would do when the pressure became unbearable".³⁶ That pressure had been gradually and deliberately raised by Russia's intense international political bullying over the issue of South Ossetia and Abkhazia. The deployment of Russian troops in Spring 2008 to Abkhazia was for Saakashvili a well sign-posted step too far and his subsequent rhetoric and actions played, Blandy believed, into Russian hands.³⁷ Two final, more recent events, have influenced Russia's development of IO; the cyber-attacks in Estonia in 2007 and the undoubted success of operation in Crimea in Ukraine in 2014 when the world met the 'little green men'.

The 2007 cyber-attack began in early 2007 when Estonian authorities decided they would relocate the 'bronze soldier', a memorial to the Red Army soldiers lost during World War II liberating the city of Tallinn from their Nazi occupiers, from its position in the center of the city to a cemetery on the outskirts of the city. The story, however, became corrupted as various Russian news sites reported, falsely, that both the statue and accompanying Soviet war graves were being destroyed. On 26th April 2007 rioting erupted in the city; one person died and over 150 were injured. On 27th April Estonia was rocked by a series of aggressive cyber-attacks which in some instances were sustained for weeks. Banks, news, and media outlets, Defence IT systems, government websites, internet service providers and small businesses were all attacked, predominantly by a Distributed Denial

---

35 *Russian Reaction to September 11 Attacks: One Extreme to the Other*. North Caucasus Weekly Volume: 2 Issue: 33, https://jamestown.org/program/russian-reaction-to-september-11-attacks-one-extreme-to-the-other-2/.
36 Blandy CW, *Provocation, deception, entrapment: The Russo-Georgian Five Day War*, UK Defence Academy ARAG paper 09/01, 2009.
37 Ibid.

of Service (DDoS). While Russia denied all knowledge of the attacks, it refused all cooperation in the ensuing investigation and observers noted that many of the attacks had either direct or indirect Russian state connections. Russia maintained that the attacks were undertaken by a grass roots activists around the world. Attacks peaked on 9th May – the hugely symbolic date that Russia celebrates its victory over Nazi Germany; they actually began at 11pm on 8th May suggesting that the attackers were using Moscow time (1 hour ahead of Estonia). The Estonian government was less circumspect and pointed at Russia almost immediately and deeming the attack a threat to national security.

How successful was the attack? Since 2007 the bronze statue has resided not in the city center but instead at the Soviet War Memorial in the Tallin Military Cemetery. But, of course, the bronze statue was only ever a subterfuge. The wider operation, which essentially shut down huge swathes of the nation for an extended period, was entirely successful. As the Estonia Cyber Defence Centre of Excellence, in its analysis of the attacks, notes: "if the cyber attacks were the result of an information operation, then one could argue that it was fairly effective. Large scale attacks were mounted against an independent state while no controlling entity (government or otherwise) has been identified".[38]

Seven years later, in Crimea, Russia flexed its muscles again. Crimea is a landmass the size of Switzerland and a heavily disputed territory since it was first annexed from the Ottoman empire in 1783 by Catherine the Great of Russia. Britain and France had demanded that Russia dismantle its Naval base in Sevastopol as part of the Treaty of Paris ending the Crimean War, but it quickly reneged and during the Franco-Prussian war in 1870 it was rebuilt. Almost completely destroyed during World War II, it was again rebuilt but in 1954 the entire Crimean region (Oblast) was transferred to the Ukrainian Soviet Socialist Republic as a gift from President Nikita Khrushchev. After the fall of the Soviet Union, Ukraine signed a 1997 treaty with Russia to allow it to keep its Black Sea fleet in Sevastopol under a lease arrangement that expired in 2042. 17 years later Russia annexed it, triggering a conflict that displaced nearly two million people. Justifying the action President Putin said that Crimea has always been an integral part of

---

38 Ottis, R. (2008). *Analysis of the 2007 Cyber Attacks against Estonia from the Information Warfare Perspective*. Proceedings of the 7th European Conference on Information Warfare and Security, Plymouth, 2008. Reading: Academic Publishing Limited, pp 163-168 available at https://ccdcoe.org/library/publications/analysis-of-the-2007-cyber-attacks-against-estonia-from-the-information-warfare-perspective/.

Russia in the hearts and minds of people and noted that the population of the peninsular was now largely ethnic Russian anyway.

The decision to seize Crimea was carefully planned. At the heart of Russia's thinking was the desire to try and avoid the need for kinetics and if they could not be avoided to ensure that whatever shooting took place was on the terms that suited them best. To do this they looked backwards to Reflexive Control and Maskirovka.[39] On 27th February, Russian troops seized the Crimean parliament building and blockaded Ukrainian military bases across the peninsular. Despite their modern Russian uniforms and weapons, the troops wore no military insignia and Moscow vehemently denied that they were Russian troops. Such an obvious and simple deception was enough to inject an uncertainty into the situation. Was it conceivable that Moscow was right, that there were in fact mercenaries or even Crimean vigilantes, and that this was some spontaneous uprising by ethnic Russia 'hotheads'?

This deliberate maskirovka, or deception operations, was enough to give the Russians and their local allies the time to take up commanding positions across Crimea, including blockading Ukrainian garrisons, such that even if they had then been ordered to fight, they would have been in a very weak position. Ultimately, they surrendered after at most the demonstrative use of a few tear gas grenades, and Russia was able to seize Crimea without a single fatal casualty.[40]

Pavel Zolotarev, a retired Major General in the Russian Army, explained, "We had come to the conclusion... that manipulation in the information sphere is a very effective tool." That conclusion was reinforced by the perception that these operations are extremely difficult to defend against, particularly with multinational military alliances like NATO, which is built to deter and, if necessary, defeat a traditional, conventional military threat. Information Warfare, in addition, is an extremely low-cost alternative to conventional military conflict.[41]

---

39 It is a term that has no direct equivalent in the West but it simultaneously encompasses the arts of: "concealment, the use of dummies and decoys, disinformation and even the execution of complex demonstration manoeuvres. Indeed anything capable of ... weakening the enemy.
40 Howard C. & Pukhov, R. (eds), *Brothers Armed: military aspects of the crisis in Ukraine*. Minneapolis: East View Press, 2014.
41 *Report of the Select Committee on Intelligence United States Senate on Russian Active Measures Campaigns and Interference in the 2016 U.S. Election. Volume 2: Russia's Use of Social Media*, available at https://www.intelligence.senate.gov/sites/default/files/documents/Report_Volume2.pdf.

The predictable confusion caused by the deployment of the 'Little Green Men' in Crimea, and the paralysis their uncertainty caused in the decision-making circles of Europe, allowed the largely peaceful annexation by Russia. Indeed, one might argue that that paralysis, which was eminently predictable, provided NATO and the West with an 'escape route'. Moscow did not want to place the West in a position where it felt ready to respond militarily – and neither did the West. Allowing itself to be 'confused' by the 'Little Green Men' suited both parties very well. That is perhaps best illustrated by the Chief of the Russian general staff, Garasimov, who wrote: "No matter what forces the enemy has, no matter how well-developed his forces and means of armed conflict may be, forms and methods for overcoming them can be found. He will always have vulnerabilities and that means that adequate means of opposing him exist. Or to put another way. You may be stronger, but we will be smarter".[42]

On 13th July 2018, a United States federal grand jury, sitting in the District of Columbia, returned an indictment against 12 Russian military intelligence officers for their alleged roles in interfering in the 2016 presidential elections. The indictment charged 11 defendants with gaining unauthorized access into computers involved in the election process, stealing, and releasing those documents to directly influence US voters. Yevgeny Prigozhin, once a close friend of President Putin reportedly told colleagues: "Gentlemen, we interfered, we are interfering, and we will interfere".[43] Prigozhin is the financial benefactor behind the Internet Research Agency which, between January 2015 and August 2017, is believed to have run more than 470 different Facebook accounts, over 50,000 Twitter (now 'X') accounts which with an output of 3.8 million Tweets provided 19 per cent of the total number of all Tweets associated with the election using the hashtags #DonaldTrump, #Trump2016, #NeverHillary and #TrumpPence16.

In August 2020, the U.S. Senate intelligence committee released its final report on Russia's efforts. The Committee concluded that: "Moscow's intent was to harm the Clinton Campaign, tarnish an expected Clinton presidential administration, help the Trump Campaign after Trump became the presumptive Republican nominee, and undermine the U.S. democratic

---

42 General Valery Gerasimov, Chief of the General Staff of the Russian Federation, writing in Military-Industrial Kurier, February 27, 2013.
43 *Russian oligarch Yevgeny Prigozhin, "Putin's chef," admits interference in U.S. elections*, CBS News, 7 November 2022, available at https://www.cbsnews.com/news/russia-us-election-interference-yevgeny-prigozhin-putin-chef-oligarch/.

process".[44] It went on to note that its impact relied upon the efforts of certain Trump campaign executives. Paul Manafort, Trump's campaign chairman, engaged with a 'Russian intelligence officer' named Konstantin Kilimnik and Russian oligarch Oleg Deripaska, with whom it said Moscow coordinates foreign influence operations. WikiLeaks published thousands of emails hacked from Clinton's campaign playing a key role in the Russian influence campaign. The Committee noted that "Trump and senior Campaign officials sought to obtain advance information about WikiLeaks's planned releases through (Republican political operative) Roger Stone".[45] Alongside Trump's staffers and Russian Intelligence Agents, were a number of other organisations who allegedly had a hand in Trump's election – amongst them the British research company Cambridge Analytica (whom I look at more closely in Chapter 11).

Of course, the question that everyone wants to know is how successful were Russian (and other) efforts? Clinton won the popular vote, taking 48 per cent of the vote compared to Trump's 46 per cent of the vote, but Trump won the electoral vote and thus the presidency. Russian influence? Cambridge Analytica? Or something domestic? Did Trump simply tap into a deep vein of US discontent, and understand the electorate, in a way that Clinton was unable to do or maybe the intervention by FBI Director Comey, just a few days before the election, which publicly reopened the Clinton emails investigation) have had an unintended but decisive effect on the election outcome?[46]

Ukraine 2022

The so called 'Special Military Operation', Putin's murderous and illegal war in Ukraine, did not begin in February 2022. Long before the full-scale military invasion, Russia had declared war on Ukraine in the information domain; not just with the troll factories and cyber-attacks but also with more respectable organisations fermenting the seeds of hate; the mainstream Russian media and, of note, the Russian Orthodox Church, were both enthusiastic amplifiers for Putin's rhetoric against the 'Nazi controlled'

---

44 Hosenball M, *Factbox: Key findings from Senate inquiry into Russian interference in 2016 U.S. election*, 18 August 2020, available at https://www.reuters.com/article/us-usa-trump-russia-senate-findings-fact-idUSKCN25E2OY.
45 Ibid.
46 Silver N, *The Comey Letter Probably Cost Clinton The Election So why won't the media admit as much?*, FiveThirtyEight.com, 3 May 2017, available at https://fivethirtyeight.com/features/the-comey-letter-probably-cost-clinton-the-election/.

lump of land to their south that was and has always been part of Mother Russia and allegedly had no right to exist as an independent state. Russia's preparation of the information battlefield was methodical and vast. As The Atlantic Council found. "for the period representing the seventy days prior to the invasion (December 16, 2021–February 24, 2022) we analysed and catalogued more than ten thousand articles by fourteen pro-Kremlin outlets. Among the primary narratives we investigated were:

- Russia is seeking peace (2,201 articles);
- Russia has a moral obligation to do something about regional security (2,086 articles);
- Ukraine is aggressive (1,888 articles);
- the West is creating tensions in the region (1,729 articles);
- Ukraine is a puppet of the West (182 articles)".[47]

Ukraine had done its best to protect its population, but it had found not just Russia but also the West to be against it. In 2016 Ukraine had sought to prohibit Russian social media inside the country but some of the loudest and most vociferous objections were from supposed western partners who lecture Kyiv on the important of Free Speech. When Ukraine became the first country to ban Russian software, for example the Kaspersky virus checker, it was again assailed by criticism from the West.

In February 2022, when Russian armed forces again crossed the border into Ukraine and drove for Kyiv some of the first targets were the huge TV towers such as the 368m tower in Donetsk. Immediately all Ukrainian TV was turned off and, in its place, Russian TV broadcast; so too the internet which was quickly routed via Moscow. I, like the Russian government, thought Ukraine would fall quickly. The sheer size and perceived capability of Russia's army mitigated heavily against Ukrainian victory. But as an MoD Advisor in Ukraine in 2014, I saw first-hand the challenges the Ukrainian political system faced and felt that too weighed against their success. But I, like the Kremlin, had reckoned without the extraordinary leadership of former actor, comedian and now, improbably, Ukrainian president Volodymr Zelensky and the astonishing resilience of the entire Ukrainian people. I write this paragraph the day after Zelensky addressed both Houses of the US Congress, in person; as Zelensky told

---

47 Carvin A (Ed), *Narrative Warfare How the Kremlin and Russian news outlets justified a war of aggression against Ukraine*, The Atlantic Council, available at https://www.atlanticcouncil.org/in-depth-research-reports/report/narrative-warfare/.

a packed chamber, "Ukraine is alive and kicking". How and when the conflict ends might, by the time this book is published, be history. In 2023 it certainly seemed that Ukraine had the advantage; in early 2024, not so much. It is a conflict that has created a great many surprises.

The first and most obvious is that the Russian Army is just not very good. The conflict has vividly shown up its poor logistics, terrible and poorly maintained equipment, lack-lustre leadership, poor tactics, lack of discipline and self-control amongst its troops and low morale. It is quite clear to me that we were again lured into a view that the Russian military was on a par with our own and we mistook mass for excellence – just as we saw when the Soviet Army fell apart post 1991. The operations in Crimea in 2014 – which was impressive – deepened that misguided belief. Setting aside the military, kinetic component, what of their IO?

In May 2022 I was invited to write a briefing document on IO in Ukraine for the UK MoD. In that short document I cautioned against too much optimism. If we were to believe social media at that time, Ukraine was winning its IO battle against Russian forces hands down. Whilst we have seen Europe and the US impose sanctions against Moscow, we had witnessed the arrival of foreign fighters into Ukraine, we had seen substantial military aid and money being donated and, overwhelming global sympathy for Kyiv, but if Ukraine had indeed 'won' then, ergo that meant that Russia must have 'lost'?

Given the volume of material, doctrine, papers, books, and commentary on hybrid warfare, on grey zone operations, on Russian troll farms, this would be a startling failure for Russia – whom we have collectively placed on an IO pedestal of excellence. But to deliberately misquote Churchill, "However beautiful the belief [sic], you should occasionally look at the results". In determining the effect of Ukrainian communication efforts, it seemed all too easy to confuse correlation and causality. I suspect much of what has happened – sympathy, money, sanctions, fighters – would have happened anyway. But, likewise, some key things the Ukrainians so desperately wanted, have not: no fly zones were not imposed by NATO; German political support has not been unequivocal; some US republican politicians are actively against supporting Ukraine. Another problem is geography. If I were sat in St Petersburg viewing VK – the Russian equivalent to Facebook – or watching mainstream Russian state TV news, I would be firmly of the view that Russia is wining an entirely justified war against a Nazi-inspired regime that has persecuted ethnic Russians inside Ukraine for years. The small-scale protests that broke out in some

Russian cities were almost certainly unrepresentative of the wider Russian population, which was either supportive of Putin or, for the vast bulk, utterly ambivalent, pre-occupied by their own battles to survive in a society ground down by years of corruption. As much as Kyiv might hope to wake to news of mutinous Russian tanks surrounding the Kremlin, the successful insulation of the Russian population for years against the west, the leveraging of the great patriotic war and the astonishing ability to paint a Russian speaking Jew, and a population with deep and historic family ties to everyday Russians, as Nazis was a massive success for Putin.

At the start of the 'Special Military Operation', Putin's principal audience was not the UK, the US or wider international groups; it was unequivocally the Russian population. Not that he was much bothered by their proactive support (although he was happy enough to manufacture it). Actually, what he wanted was their inactivity; inactivity for Putin = freedom of manoeuvre. In that respect we can, with certainty, see a causal relationship between his control of the [Russian] information space and his freedom of manoeuvre. Putin spent years preparing his cognitive battle space. He may not have enjoyed the success he anticipated in former soviet states, but he has certainly achieved his objectives in Russia.

With the domestic Russian audience captured, Putin's influence machine has had room to work on others – and at the top of his list is Ukraine's biggest military donor – the United States. With little sympathy from the US President and incumbent democrats, Putin has instead focussed attention on Republicans, many of whom remain loyal to Trump. In December 2022 CNN published an article on what it called declining Republican Party support: "The level of support among Republicans for the US giving military aid has also declined: 55% were found to be in favour compared to 68% in July and 80% in March".[48] Various investigations have revealed close connections between the Republican party and Russian oligarchs with close ties to Putin.[49] North Carolina congressman, Madison Cawthorn, has said that: "the Ukrainian government is incredibly corrupt, and it is incredibly evil, and it has been pushing woke ideologies", whilst Republican congresswoman, Marjorie Taylor Greene, appeared on

---

48 Britzky H, *Poll finds Republican support for US aiding Ukraine's war effort against Russia is declining*, CNN, 6 December 2022, available at https://edition.cnn.com/2022/12/06/politics/us-support-ukraine-aid-russia-poll/index.html.

49 Friedman D, *Russians Used a US Firm to Funnel Funds to GOP in 2018. Dems Say the FEC Let Them Get Away With It*, 30 October 2022, available at https://www.motherjones.com/politics/2022/10/russians-used-a-us-firm-to-funnel-funds-to-gop-in-2018-dems-say-the-fec-let-them-get-away-with-it/.

conservative radio show 'Voice of Rural America' and defended Putin: "You see, Ukraine just kept poking the bear, and poking the bear, which is Russia, and Russia invaded".[50] Fox TV host, Tucker Carlson – a hugely influential voice in the MAGA and Trump wing of the Republican Party – asked the media: "Why shouldn't I root for Russia, which I am?"[51] All of which explains why President Zelensky's first overseas visit was to the US. Although he received the expected standing ovation not all were supportive. Warren Davidson, republican member for Ohio, was asked if he would support further funding: "no chance" was his reply. Florida Republican, Matt Getz, reflected on Zelensky's speech: "He did not change my stance on suspending aid for Ukraine",[52] whilst Marjorie Taylor Green – another republican – declared that the US was waging a 'proxy war' on Russia.[53] Ukraine based Euromaiden Press ran an article by Anton Shekhovtsov, in June 2023 looking at the specific Russian narratives by intended audience. For the US, it notes two in particular: "Ukraine is run by Nazis" and "Ukraine is one of the most corrupt countries in the world – it cannot be part of the West".[54]

In December 2022 the New York Times reported on leaked Russian emails detailing how a narrative of Russia winning in Ukraine was crafted for Russian audiences from American and Chinese news.[55] The paper reported that: "Russian propagandists plucked clips from American cable news, right-wing social media and Chinese officials. They latched onto claims that Western embargoes of Russian oil would be self-defeating, suggested that the United States was hiding secret bioweapon research labs in Ukraine and that China was a loyal ally against a fragmenting West".[56] A particular 'favourite' are clips of the Fox News host, Tucker Carlson, a controversial broadcaster and keen supporter of Donald Trump who

---

50 Fedor L, *Pro-Putin Republicans break ranks by heaping praise on Kremlin*, The Financial Times, 26 March 2022, available at https://www.ft.com/content/fd870fa9-007a-4cd4-bffc-d72aa2a35767.
51 Rupar A, *Tucker Carlson's defense of Russia takes "America First" to its logical conclusion*, VOX, 26 November 2019, available at https://www.vox.com/2019/11/26/20983778/tucker-carlson-rooting-for-russia-ukraine-invasion-america-first.
52 Ibid.
53 Ibid.
54 Shekhovtstov A, *Four towers of Kremlin propaganda: Russia, Ukraine, South, West*, Euromaiden Press, 6 January 2023, available at https://euromaidanpress.com/2023/01/06/russian-propaganda-war-related-strategic-and-tactical-narratives-and-their-audiences/.
55 Mozur P, *An Alternate Reality: How Russia's State TV Spins the Ukraine War*, New York Times, 15 December 2022, available at https://www.nytimes.com/2022/12/15/technology/russia-state-tv-ukraine-war.html.
56 Ibid.

claimed the 2020 US election had been rigged.⁵⁷ "Be sure to take Tucker," one Russian news producer wrote to a colleague in an email seen by The New York Times. Carlson's commentaries on the Ukraine war has, in the view of some, "generally reflected Putin's speeches and claims. Russian television then plays back the monologues as evidence that Putin is right because the same is being said by 'the most popular television presenter in America'".⁵⁸

In December 2022, General Mark Milley, the US Chairman of the Joint Chiefs of Staff, estimated that Russian deaths and serious casualties now exceed 100,000.⁵⁹ The 11 months of war had seen a 560 per cent increase in casualties on the total number in the Soviet Union's ten-year war in Afghanistan. And whilst 300,000 young men may have left the country to avoid the draft, there are still over 12 million males aged 20-40 to throw into the fight. But more startling, despite the clear incompetence of the draft, the obvious failure of the 'special military operation' and the staggering costs, Russians are either rallying around the flag (a smaller number) or remain stoically ambivalent (a much larger number). Helping that is Putin's tightly controlled media machine.

Whilst the international community may not be Putin's principal audience that is not to say he has ignored them. On the 13th of October British Member of Parliament, Tobias Ellwood, tweeted "Global isolation for Putin"⁶⁰ as 143 nations vote against Russia at the UN General Assembly. What he failed to note was that 35 nations abstained, including India, South Africa, Pakistan, Sri Lanka (all Commonwealth nations) and of course, China. In April 2023 the Egyptian governor was forced to deny that, despite international sanctions, it had been planning to sell arms to Russia.⁶¹ And in a conversation with a friend and senior advisor in the Abu Dhabi government, I was told: "…this is a European problem, not our one. And

---

57 Baragona J, *Tucker Breaks into Fox News Election Night Coverage to Sow Doubt*, The Daily Beast, 9 November 2022, available at https://www.thedailybeast.com/tucker-carlson-breaks-into-fox-news-election-night-coverage-to-sow-doubt.
58 McGreal C, *Who is Tucker Carlson really 'rooting for' in Ukraine?*, The Guardian, 2 October 2022, available at https://www.theguardian.com/media/2022/oct/02/tucker-carlson-ukraine-vladimir-putin-propaganda.
59 *Ukraine war: US estimates 200,000 military casualties on all sides*, BBC News, 10 October 2022, available at https://www.bbc.co.uk/news/world-europe-63580372.
60 Twitter Account @Tobias_Ellwood Tweet dated 13 October 22.
61 Staff T, *Egypt planned to supply thousands of rockets to Russia amid Ukraine war – report*, Times of Israel, 11 April 2023, available at https://www-timesofisrael-com.cdn.ampproject.org/c/s/www.timesofisrael.com/egypt-planned-to-supply-thousands-of-rockets-to-russia-amid-ukraine-war-report/amp/.

anyway, the US and UK have not been reliable partners for GCC nations. They expect free support. They won't get it". As if to underline that the very next day the Kremlin posted the arrival of UAE president Sheikh Mohamed bin Zayed al Nahyan to Moscow on social media. Sheikh Mohamed spoke warmly about the increase of trade with Russia from US$2.5 to US$5 billion; he spoke of the 4,000 Russian companies now based in United Arab Emirates and the opening of the first Russian school. What he did not talk about (at least in public) was Ukraine. Whilst the decision of Sweden and Finland to join NATO is hugely symbolic, so too Putin's indictment at the International Criminal Court,[62] global support cannot be taken for granted. Perhaps with this in mind President Zelensky travelled to the US in December 2022 to address both Houses of Congress, a move timed to pre-empt the Republican party taking control of the House in January 2023. In a survey for Politico in November 2022, 48 per cent of Republicans said the US was doing too much for Ukraine, compared to 6 per cent in March[63] and a growing number of it seems extraordinary that voices should be supporting Putin, but to some Putin is clearly 'their man' – and stopping the flow of US weapons and aid to the "Nazi regime" in Kyiv (conveniently and seemingly without any contradiction having a Jew – the enduring enemy of the right – at its head) a loudly vocalised objective. The 'Nazi' reference, so freely used by Putin to describe Ukraine, was much in evidence on Twitter (now 'X') on the day of Zelensky's visit; @dana916 tweeted of the 'shame' of US flags alongside Ukrainian ones; @Birrion asked if Zelensky was "bringing his Nazi bodyguard to Washington", a later tweet reminding his 4,000 followers that "the Maidan uprising was a US backed coup which normalised Nazism"; @realcallumB tweeted that "the Nazi criminal [Zelensky] meets the incompetent criminal [Biden]" and @scotty2971 told his 6,000 followers that "Zelensky was here in Washington while his military his killing his own people. He is a Fucking Nazi". Sadly, there are hundreds and possible thousands of such tweets. Many will be Russian in origin, or automated bots, but many will not. Only with large scale data analysis can a true sense of the Twitter (now 'X') sphere be assessed but even then, with so many fake and automated accounts it would likely be a useless metric.

---

62 Borger J, *ICC judges issue arrest warrant for Vladimir Putin over alleged war crimes*, The Guardian, 17 March 2023, available at https://www.theguardian.com/world/2023/mar/17/vladimir-putin-arrest-warrant-ukraine-war-crimes.
63 Allen M, *Republicans Don't Have to Write a 'Blank Check' for Ukraine*, Politico Opinion, 17 November 2022, available at https://www.politico.com/news/magazine/2022/11/17/republicans-dont-have-to-write-a-blank-check-for-ukraine-00069260.

Twitter (now 'X') has proven an important battle ground in the war for Ukraine. Whilst its government has focussed on the strategic communication objectives, the daily grind of war 'reporting' has been taken up by dispersed units and others inside the country. They have proven adept at using it to showcase operations against Russian units with huge numbers of accounts tweeting drone and front-line footage of Russian soldiers being targeted and killed, often to heavy metal music tracks. The output is significant and mirrored across other social media sites such as TikTok and Instagram. Dean of the School of Communications at the University of Miami, Karin Wilkins, told News@TheU that: "The internet has also allowed interaction between Ukrainians and people across the globe. These exchanges have helped the Ukrainians feel less isolated and have empowered many to reveal their personal stories". Heidi Carr, assistant professor of professional practice at the school said: "I can't think of any war, not even in the 21st century wars in Iraq and Afghanistan, where social media has had such an impact".[64]

Ukraine may well be wining the tactical and operational information campaign – although not amongst Russian audiences which given the historic connectivity between the two nations and the fact that huge numbers of Ukrainians live in Russia, or are married to Russians, or have historical family ties to Russia, is odd. However, Russia in 2024 is certainly not losing the strategic information campaign. That may change; the accession of Finland into NATO is significant. So too Putin's indictment at the International Criminal Court. The invasion may have prompted what the Japanese Foreign Ministry described as an 'historical inflection point'[65] but Russia has still been able to take over the chairmanship of the United Nations Security Council in April 2023.[66]

But that may not be a dynamic that is sustainable, as the Centre for European Policy Analysis published in April 2023. "Putin enjoys the support of a population that does not trust him, but that is learning to

---

64 Gutierrez B, *Invasion of Ukraine highlights Social Media's role in war*, News@theU, 9 March 2022, available at https://news.miami.edu/stories/2022/03/invasion-of-ukraine-highlights-social-medias-role-in-war.html.
65 *Russian war in Ukraine shows end of post-Cold War era, Japan says in report*, The Japan Times, 11 April 2023, available at https://www.japantimes.co.jp/news/2023/04/11/national/politics-diplomacy/russian-war-ukraine-shows-end-post-cold-war-era-japan-says-report/.
66 Borger J, *Absurdity to a new level' as Russia takes charge of UN security council*, 31 March 2023, The Guardian, available at https://www.theguardian.com/world/2023/mar/31/absurdity-to-a-new-level-as-russia-takes-charge-of-un-security-council.

trust one another even less. For the time being, the balance of that equation works out in Putin's favor: The best way for Russians to protect themselves against their fellow citizens is to make it clear that they support Putin and the war. The longer the war lasts, however, the riskier that strategy becomes".[67]

The long awaited Ukrainian counterattack of 2023 seems not to have been successful as was hoped by the international community. That likely had some impact in the US, in particular, where President Biden's request for further financial and military aid has become victim to the primacy of domestic politics. That is not to say there have not been significant successes as well – the US decision to allow European nations to release F16 aircraft to Ukraine being an obvious example. Perhaps an indisputable consequence of the Ukraine war – even a success – is that it has at last galvanised an until now largely complacent West into looking more seriously at IO; maybe finally leaders and policy makers are waking up the fact that IO don't just support conventional military operations from the periphery, today they are at the epi-centre of conventional military operations.

---

[67] Greene S, *The Russian Rally*, 7 April 2023, available at https://cepa.org/article/the-russian-rally/.

8 Iran, North Korea and China

I had the privilege of working alongside one of the UK's finest Iranian analysts whilst working in the MoD's Advanced Research and Assessment Group in 2009. We kept in occasional contact and one day he emailed to tell me that my name had suddenly 'pinged' on the radar of the Iranian regime that he so closely monitored. Slightly perturbed I asked for more details and he told me, with a little more amusement in his voice than I thought strictly necessary, that Ayatollah Mohammad Taqi Mesbah-Yazdi, one of the most conservative and hard-line of the regime's clerics, could not let go of the name, my name, '*Estiv Taataam*', apparently having come across it after I spoke at a conference in the Middle East. It was slightly disconcerting; as a published and vocal supporter of military IO officer I'd long accepted that Moscow probably had a file on me, but Tehran? Better the devil you know than one you do not! And whilst we know plenty about Russia's IO, we know a lot less about the Islamic Republic of Iran.

Iran, which is almost three times the size of France or slightly smaller than the US state of Alaska, is home to over 88 million people, the vast bulk of whom are aged under 40 years old. Home to one of the world's oldest continuous civilisations; 500 years before Christ it was the world's first superpower, the Achaemenid Empire, ruling from eastern Europe, across north Africa and into Asia.[1] Winston Churchill, in a 1939 radio broadcast, defined Russia as "a riddle, wrapped in a mystery, inside an enigma".[2] The description could just as equally apply to Iran.

Iran is the centre of Shia Islam (when the Prophet Mohamed died in 632, they believed that his son in law, Ali, was his rightful successor. Sunni Muslims do not recognise Ali and the theological split has existed ever since) which has been in ideological (and often proxy led) conflict with Sunni Islam ever since. Whilst its President is elected democratically, so too the Parliamentary members, there the resemblance to a western style

---

1 Holland, T, *Persian Fire: The First World Empire*, Battle for the West. Hachette, 2011, ISBN 0748131035.
2 See https://www.oxfordreference.com/display/10.1093/acref/9780199567454.001.0001/acref-9780199567454-e-1570.

democracy ends and ultimate control in the country sits with the Supreme Leader – the highest political and religious authority. Since 1989 the Supreme Leader has been Ali Hosseini Khamenei, now the longest serving head of state in the middle east. Head of State and Commander-In-Chief of the Armed Forces, almost all of Iran's decision making is vested in him. He, and Iran's political system are protected by the IRGC – the Islamic Revolutionary Guard Corps – an ideologically driven military elite that exists alongside but separate from the country's conventional armed forces. Within the IRGC exists the Quds Force, a unit responsible for clandestine operations which until 3rd January 2020 was commanded by Major General Qassim Soleimani; at 1am, near Baghdad Airport in Iraq, Soleimani was assassinated by a US drone strike.[3]

Iran's IO are directed both internally to its own population and externally in direct and focussed support of its strategic objectives. Iran is not particularly interested in anything that does not meet either of these two aims. The US elections were an obvious target; aside from the long-standing enmity between the two countries, Iran believes that the US is an existential threat to its existence. The US is also close friends with its mortal ally, Israel, and it re-imposed sanctions after pulling out of the Joint Comprehensive Plan of Action (JCPOA – the arrangement brokered by China, France, Germany, Russia, the United Kingdom and the United States) in 2015 to limit the Iranian nuclear programme. The UK, as a staunch US ally, has also been a target for Iranian activities. For example, since 1979 Iranian leaders have been interested in exploiting the conflict in Northern Ireland and the emergence of the Scottish independence movement to promote separatism. In the early years of the post-revolutionary period, Iranian hard-liners publicly supported the Irish Republican Army (IRA) terrorists and even changed the name of a road in Tehran from Churchill Street to Bobby Sands[4] Street. In a rather macabre gesture, some Iranians who supported IRA terrorists, opened a burger shop and named it after Bobby Sands, a convicted IRA terrorist who had died while on hunger strike in prison.[5,6]

---

3 BBC News, *Qasem Soleimani: US kills top Iranian general in Baghdad air strike*, 3 January 2020, available at https://www.bbc.co.uk/news/world-middle-east-50979463.
4 Bobby Sands was a member of the Irish Republican Army and, latterly, a Member of Parliament, who died whilst on hunger strike in May 1981.
5 *Two spots in Tehran, Iran pay homage to Bobby Sands*, Irish Central, 13 January 2020, available at https://www.irishcentral.com/roots/iran-bobby-sands.
6 Costello N, *The Capital of Iran Has a Burger Joint Dedicated to an IRA Hunger Striker*, Vice, 16 January 2015, available at https://www.vice.com/en/article/9bzpya/bobby-sands-burgers-tehran-545.

More recently, the escalation of the nuclear crisis and the deterioration of Anglo-Iranian relations led some Iranian hard-liners to support the Scottish Nationalist Party (SNP). Former SNP leader Alex Salmond visited Iran as the head of an SNP delegation in 2015 and held trade talks with Iranian officials. One MSP described the visit as "hugely positive". Salmond even went so far as to declare that "Iran could be a valuable ally" for the SNP.[7] Thus the Scottish independence referendum of 2014 became an issue of some interest to Tehran. Iranian cyber warriors associated with state media used fake social media accounts to interfere in the referendum debate, including creating fake social media accounts that looked like the Scottish national newspaper 'The Scotsman' and using it to spread cartoons attacking the then Prime Minister, David Cameron, portraying him as an agent of English oppression.[8,9] However, like so much about Iran, the matter is by no means clear cut. Iranian President Rouhani studied for a master's degree in law at Scotland's Glasgow Caledonian University, graduating in 1999.[10] He is not on public record as having spoken in favour of Scottish independence, which given his first-hand knowledge might have been expected, however it is likely that others in the regime believe it can be exploited for the purpose of gaining policy leverage over the UK.

Iranian leaders have tried to interfere into UK politics more directly. During the riots that spread across the UK between 6-11th August 2011, following the shooting of Mark Dugan by the Police,[11] Iran accused the UK of "double standards on the human rights issue", condemned its "police savagery",[12] and called on the British authorities to "exercise restraint". Moreover, the then commander of the Basij Resistance Force of the IRGC, Mohammad Reza Naqdi, went so far as to declare: "Should the General Assembly of the United Nations approve, the Basij is ready to send some of its Ashura and Al-Zahra battalions to London, Liverpool, and Birmingham

---

7 *Alex Salmond: Iran could be valuable ally*, The Courier, 28 December 2015, available at https://www.thecourier.co.uk/opinion/244886/alex-salmond-iran-could-be-valuable-ally/.
8 Hamil J, *Iranian hackers tried to help Scotland become independent, Facebook claims*, The Herald, 6 May 2020, available at https://www.heraldscotland.com/news/18430151.iranian-hackers-tried-help-scotland-become-independent-facebook-claims/.
9 Nimmo B et al, *Iran's Broadcaster: Inauthentic Behaviour*, Graphika, May 2020, available at https://publicassets.graphika.com/reports/graphika_report_irib_takedown.pdf.
10 ویدیوی مظحل کـردم ارتکـد طسـوت نـحن روحانی در لاس ۱۹۹ Elections in Iran YouTube Channel, available at https://www.youtube.com/watch?v=dkKaSIRwiJw.
11 The Killing of Mark Duggan, Forensic Architecture.Org, available at https://forensic-architecture.org/investigation/the-killing-of-mark-duggan.
12 *Iranians protest over UK police "savagery" in riots*, Reuters, 14 November 2011, available at https://www.reuters.com/article/us-iran-britain-protest-idUSTRE77D20O20111814.

as peacekeepers and as a buffer between the people of England and the oppressive royal regime".[13]

Internally its IO are as likely to be directed against the Iranian President (if it is perceived in anyway to be endangering or pressuring the Supreme Leader) and suppressing internal unrest as they are against the US or Saudi Arabia. Former Iranian intelligence chief, Heidar Moslehi, is on record as saying that, "We do not have a physical war with the enemy, but we are engaged in heavy information warfare with the enemy"[14] and that enemy is very clearly western culture values, the West's foreign policy and internal instability. For the regime any loss of control of information is an existential threat: "Should the information conflict be lost, many Iranian officials believe the collapse of the state will soon follow".[15] Finally, Iran uses IO as a form of religious lobbying. In his book Religious Statecraft, Professor Mohammad Ayatollahi Tabaar, who initially studied in Iran before relocating to the US, explains how Iran develops and deploys Shi'a Islam inspired ideologies to: "gain credibility, constrain political rivals, and raise mass support".[16]

Domestically the regime has prioritized information control over its population – of whom it is estimated that nearly 60 million are internet users.[17] The 2009 Green Movement, the name given to the spontaneous mass demonstrations that erupted against the officially declared Presidential victory of Mahmoud Ahmadinejad, and which continued until 14th February 2010 when a protestor's rally was brutally supressed and the movements leaders arrested and imprisoned, was largely attributed to the use of social media and the regime cracked down hard. As well as taking real accounts offline, Iran began operating a huge number of so called 'sock puppet', or faked, accounts on Facebook and Twitter (now 'X"). By February 2020 Facebook had identified in excess of 2200 such accounts whilst Twitter (now 'X') had taken down over 8000 accounts.[18]

---

13 *Basij Commander Joins Tehran Chorus Lampooning U.K. over Riots*, Radio Free Europe, 12 November 2011, available at https://www.rferl.org/a/iran_basij_commander_uk_riots/24295464.html.
14 *Iran Adopts Aggressive Approach Toward Enemies*, Tehran Times, 18 July 2011, available at https://www.tehrantimes.com/news/244231/Iran-adopts-aggres- sive-approach-toward-enemies.
15 Brooking & Kianpour, *Iranian digital influence efforts: Guerrilla broadcasting for the twenty-first century*. The Atlantic Council, 11 February 2020, available at https://www.atlanticcouncil.org/in-depth-research-reports/report/iranian-digital-influence-efforts-guerrilla-broadcasting-for-the-twenty-first-century/#Chapter_1.
16 Tabaar, *Religious Statecraft. The Politics of Islam in Iran*. Colombia University Press, 2019.
17 Brooking & Kianpour, op.cit.
18 Ibid.

Buoyed by their apparent success in the domestic arena, Iran has extended its operations internationally becoming one of the largest state based global practitioners of IO. As the US based Atlantic Council have reported, "Iran makes less use of obvious falsehood. Instead, Iran advances a distorted truth: one that exaggerates Iran's moral authority while minimizing Iran's repression of its citizens and the steep human cost of its own imperial adventures in the wider Middle East".[19]

In March 2021, the US' National Intelligence Council released a report entitled *'Foreign Threats to the 2020 US elections'*.[20] They offered five key judgments about what interference had, and had not taken place, and in judgement three determined that Iran had carried out a: "multi-pronged covert influence campaign intended to undercut President Trump's re-election prospects ... undermine public confidence ... and sow division and exacerbate societal tensions in the US". What did that interference look like?

On 20th October 2020, Iran sent out a series of faked emails to US voters. The emails purported to be from the so called 'Proud Boys'. The Proud Boys claim they are no more than a loose group of friends and associates fighting political correctness and, in the wake of the Black Lives Matter campaign, so called anti-white guilt. However, the group, whose membership is only male, was officially classified in 2018 by the US Federal Bureau of Investigation (FBI) as an extremist group with a misogynistic, islamophobic, transphobic and anti-immigration agenda. The Iranian emails threatened US voters that the Proud Boys would come after them if they did not vote for Donald Trump. The intent of the IO was to directly associate Trump (who only a week before had imposed significant new sanctions on Iran) with right wing extremists. Retrospectively this rather small-scale operation appears almost humorously naive; President Trump probably needed little 'help' in this regard – his own words often portrayed his support for hard right politics and groups.

A much larger, previous, operation was uncovered in an investigation by the Reuters news agency and the cyber security providers, Fire Eye, in 2018. They concluded that Tehran has established a global network of news sites and multiple associated social media accounts reaching into the US, the UK, Latin America, and the Middle East. These news sites, purporting to be

---
19 Ibid.
20 US Government, *Foreign Threats to the 2020 US Federal Elections*, DNI, 10 March 2021, available at https://www.dni.gov/files/ODNI/documents/assessments/ICA-declass-16MAR21.pdf.

organic, disseminated materials supporting Iranian national interests and the content was largely anti-Saudi, anti-Israeli, and pro-Palestinian. They exposed outlets in the US such as Liberty Front Press, purportedly located in San Jose in California, the US Journal registered in Palo Alto in California and the Real Progressive Front registered in Houston in Texas. Each site had associated social media channels and various named individuals. For example, the US Journal features Elizabeth Tacher (twitter user @BethTacher) as one of its writers; no such individual appears to exist and the profile pictures are stolen from a French Actresses Facebook page.[21] However, some genuine individuals were fooled enough by the deception to have their (real) names associated with the site and some used that association in their own personal resumes and marketing – an Iranian equivalent to the Russian 'useful idiot' principle which we saw in Chapter 5.

In the UK, the British Left site described itself as "a non-governmental news organization completely independent of any advertisers, funders, companies, political organisations, or political parties" and was registered in Sheffield in Yorkshire. Critics Chronicle – apparently an independent news outlet based in Birmingham – is also linked to an Iranian mobile phone number. Like their US counterparts the UK sites have their own staff writers – for example, Elena Kowalsky from Liverpool who, like Ms Tacher, uses a profile picture from someone else and has a Twitter (now 'X") account linked to an Iranian country code mobile phone. Similar sites were set up in Egypt and in South America, all with their own networks of writers and social media influencers.

The US based network was largely taken down in late 2020 by the US Department of Justice[22] however by then many of them had been operating for several years. Almost every single one of them proactively tried to hide their connections to Tehran and almost all claimed that they were independent news sites, free of any external funding or influence. The reality of course is a little different. Whilst a lot of the site content was taken from genuine news providers – and either doctored or misrepresented – a large number of stories were generated by Tehran's own version of the Russian Internet Research Agency, the International Union of Virtual Media (IUVM) and the Islamic Radios and Televisions Union (IRTVU).

---

21 *Suspected Iranian Influence Operation*, Fireeye, 2018, available at https://www.fireeye.com/content/dam/fireeye-www/current-threats/pdfs/rpt-FireEye-Iranian-IO.pdf.
22 US Government, *United States Seizes Domain Names Used by Iran's Islamic Revolutionary Guard Corps*, US Justice Department Press Release, 7 October 2020, available at https://www.justice.gov/opa/pr/united-states-seizes-domain-names-used-iran-s-islamic-revolutionary-guard-corps.

Both of these organisations have been sanctioned by the US Department of the Treasury[23] because of their direct links to the Islamic Revolutionary Guard Corps.

The global COVID-19 pandemic provided several nations with an excellent opportunity to bolster anti-US rhetoric with many pushing the idea that the coronavirus was actually a US biological weapon. On 13th March 2020, IVUM quoted Ayatollah Khamenei as saying that: "some evidence [exists] that this incident might be a biological attack" and referenced a statement made by the Chinese Foreign Ministry that the US Army had deployed Covid to Wuhan. Ten days later IVUM published an article asking if coronavirus was "part of America's biological warfare"? The article wrote that: "it is no coincidence that the virus selectively goes to countries that are considered enemies of the United States... this does not seem to be a coincidence".[24] As we will see in a later chapter, this linking of a virus to the US military is not new. Throughout the 1980s the then Soviet Union actively claimed through its proxies that AIDS had been a US military biological weapon. The Iranian coronavirus message, broadcast across the full range of covert and overt channels was picked up and repeated across the Middle East. In a study undertaken by USCENTCOM in March 2020, it was found to have influenced people's perceptions in Iraq, north east Syria, Lebanon and in Yemen, where the Houthi leader Abdulmalik al-Houthi described COVID-19 as a biological weapon released by the US, Saudi Arabia and the United Arab Emirates.[25] Pro-Iranian social media amplified the message across Arabic language Twitter (now 'X') using hashtags #Bases_of_the_Amercian_Pandemic and #Coronavirus_is_Trump's_weapon.

Iran has been quick also to capitalise on the Washington Riots of 6th January 2021, running both Twitter (now 'X') and Telegram campaigns with pictures of the U.S. Capitol building in flames. Much of their invective was again aimed at President Trump whom they held responsible for the violence and whom they claimed would be responsible for ultimately destroying America.

---

23 *U.S. sanctions three new Iranian organizations for alleged election disinformation*, Reuters, 22 October 2020, available at https://www.reuters.com/article/usa-election-security-sanctions-int/u-s-sanctions-three-new-iranian-organizations-for-alleged-election-disinformation-idUSKBN277346.
24 *Iran's IVUM turns to Coronavirus*, Graphika, April 2020, available at https://public.assets.graphika.com/reports/Graphika_Report_IUVM_Turns_to_Coronavirus.pdf.
25 Whiskeyman A, Berger M, *Axis of Disinformation: Propaganda from Iran, Russia, and China on COVID-19*, Fikra Forum, 15 Feb 2021, available at https://www.washingtoninstitute.org/policy-analysis/axis-disinformation-propaganda-iran-russia-and-china-covid-19.

As well as their covert or disguised IO, Iran also has its more overt media organisations – their equivalent to Russia's RT and Sputnik. For Arabic speakers Iran maintains the Al-Alam TV channels, widely available across the Middle East; for Spanish speakers it offers HisPan TV and for the English-speaking world, Press TV. Since they are openly acknowledged as Iranian government run channels it is perhaps fairer to classify them as public diplomacy tools rather than IO and they have opened their doors to a range of views. For example, the former UK Labour Party Leader, Jeremy Corbyn, has hosted a show on the Press TV channel, so too Members of Parliament from the right and centre ground of UK politics. But like RT and Sputnik, Press TV has not been averse to amplifying news of a more dubious origin. It was, for example, accused of faking reports over US drone strikes[26] and in January 2012 the channel had its UK broadcast license revoked after airing an interview with imprisoned Newsweek presenter, Maziar Bahari, which had been conducted under duress, and latterly for failing to comply with UK editorial standards. Press TV now only maintains its UK presence via its website.

Much less is known about the Iranian armed forces use of IO. In 2010 a special military force, the Unit of the Soft War, was established, with responsibilities for so-called 'soft' operations such as propaganda and Psychological Operations.[27] The US Defence Intelligence Agency (DIA) assesses that since 2014, Iranian military leaders have been pushing for a broadening in their military capability to address a wider range of 'unconventional' threats.[28] An interesting dimension to Iranian IO is that alongside conventional doctrine that we in the west would recognise, Iran also needs to find religious justification. In a 2018 paper, Psychological Warfare is justified by references to the Holy Quran, Islamic history and the 'consensus of scholars'.[29]

---

26 *Iranian Press TV accused of faking drone strike reports*, The Telegraph, 2 November 2011, available at https://www.telegraph.co.uk/news/worldnews/middleeast/iran/8932829/Iranian-Press-TV-accused-of-faking-drone-strike-reports.html.
27 Adelkhah N, *Iran Integrates the Concept of the "Soft War" into its Strategic Planning*, Jamestown Monitor, Terrorism Monitor Volume: 8 Issue: 23, 2010, available at https://jamestown.org/program/iran-integrates-the-concept-of-the-soft-war-into-its-strategic-planning/.
28 https://www.dia.mil/Portals/27/Documents/News/Military%20Power%20Publications/Iran_Military_Power_LR.pdf.
29 Chavoshi & Ahmadi, *The jurisprudential study of psychological warfare from imam's point of view*, January 2018 Opcion 34(14): 1448-1477, available at https://www.researchgate.net/publication/329671408_The_jurisprudential_study_of_psychological_warfare_from_imam%27s_point_of_view.

After the 2010 Stuxnet cyberattack on Iranian centrifuges, Iran significantly invested in their cyber capabilities and their offensive cyber-attacks against the UK and US are now well known. In the UK a 2018 attack on the Post Office and some local government networks saw thousands of personal details being stolen; a similar attack in 2017 was launched by Iran on the UK's parliamentary IT system resulting in the loss of over 10,000 data records.[30] In the US the campaign had been far deeper and longer. At least nine major cyber-attacks are attributed to Iran including taking control of the operating system of the Bowman Dam in New York in 2013, a ransomware attack on Atlanta's city government in 2018 and an alleged hack of the US presidential campaign in late 2019.[31]

Iran has clearly advanced its IO capability over the last ten years and through both overt and covert means it is not just physically targeting western computer systems but actively sowing disinformation in strategic audiences globally. How the US government chooses to engage with Iran going forward is likely to affect how and where Iran flexes its IO muscles, but Iran has now added a very powerful tool to its armoury and has no compunction in using it if it believes it will enhance its national objectives.

The Democratic People's Republic of North Korea (DRPK)

North Korea is often referred to as the Hermit Kingdom; that is very far from the truth. Hermits traditionally live in peace and solitude, have little to no interaction with anyone and often the term is applied to those with very strict religious views. North Korea is the anthesis of these. It lives not in solitude but in the glare of the world's media – which it does much to promote – and its interactions with other nations are substantial and sustained. If there is one possible similarity it is the reference to religion. Its leadership, a hereditary dictatorship, fosters a god like image amongst its people who must genuflect before portraits and statues of the Great Leader (Kim Il Sung), the Dear Leader (Kim Jong-Il) or the current premier, the Dear Respected Leader, Kim Jong-Un. To not take North Korea seriously is a mistake. Seoul, the capital of neighbouring South Korea has a population of nearly 10 million people and they collectively live less than 60km from

---

30 Bunkall A, *Iran conducted 'major cyber assault' on key UK infrastructure*, Sky News, 3 April 2019, available at https://news.sky.com/story/iran-conducted-major-cyber-assault-on-key-uk-infrastructure-11676686.
31 *Cyber Security Awareness Training that Employees Love to Take*, available at https://www.metacompliance.com/blog/irans-cyber-attack-timeline-2009-2020/.

the Demarcation Zone (border) between democratic, capitalistic, South, and dictatorial, totalitarian North. North Korea has over 1.1 million personnel in uniform and an armoury of missiles that might soon have reach to continental United States.

Retired US Army Colonel David Maxwell has spent a lifetime serving on, and studying, the Korean peninsula and in his view "Information is critical to everything in North Korea – because information is at the heart of control".[32] He talks about the many discussions he has had with North Korean defectors and how they take a very long time to be 'deprogrammed'. One of them told him that all North Koreans live in a state of 'psychological paralysis' (what we might better refer to as cognitive dissonance) in that they know the truth of the outside world, but they have to set that aside and perform all of the rituals and duties that the North Korean state demands. Maxwell recalls that one of them told him that paralysis requires everyone to be liars in order to live. They have to lie about themselves; they have to lie to their wives, their children, their neighbours. They have to lie every day in order to just exist for fear of betrayal. And it is that paralysis, reinforced by information, that stops the population from rising up against what President Donald Trump referred to as the 'rocketman'.[33]

Information control is exerted by a number of different organisations inside Pyongyang. The first is the so-called Worker's Party of Korea's Propaganda and Agitation Department (PAD) which has three objectives. Firstly, to reinforce the reputation of the dear respected leaders and the Kim dynasty amongst both the internal North Korean population and international target audiences. Secondly, the undermining, at every opportunity, of the legitimacy of the Seoul government and thirdly, to reduce and counter the US presence on the Korean peninsula. The PAD, like most major organisations of the DPRK government, is presided over by a close Kim family member, in this case Kim Yo Jong,[34] Kin Yong Un's younger sister, who has been touted as a potential successor to the current

---

32 Interview Tatham / Maxwell Senior Fellow Foundation for Defense of Democracies, Zoom, 11 March 2021.
33 Vallejo J, *Trump obsessed with sending CD of Elton John's 'Rocket Man' to Kim Jong-un, ex-aide Bolton claims*, The Independent, 17 June 2020, available at https://www.independent.co.uk/news/world/americas/us-politics/trump-rocketman-cd-kim-jong-un-elton-john-john-bolton-book-a9572056.html.
34 Madden M, *North Korea's New Propagandist?*, 38North, 14 August 2015, available at https://www.38north.org/2015/08/mmadden081415/.

leadership.³⁵ Alongside the PAD is the Reconnaissance General Bureau (RGB) which looks after intelligence and special operations. Within the RGB is the infamous Bureau 121 which is responsible for North Korea's cyber-attacks. The United Front Department is yet another department which works on information control and domination; the UFD is particularly focussed on reaching out to, and subverting, foreign Non-Governmental Organisations (NGO).

The PAD's job is not an easy one. Conventional wisdom is that North Korea is completely isolated from the rest of the world; conventional wisdom is wrong. Research by the Centre for Strategic and International Studies found that almost 92 per cent of respondents (interviewed inside North Korea) were able to consume foreign media at least once a month and 83 per cent assessed that foreign media had a greater interest and impact on their lives than the decisions made by the North Korean Government.³⁶ That foreign media enters the country in two specific ways – via cross border smugglers on DVDs and USB sticks (or indeed hung beneath balloons that are infrequently floated across the border) and via the increasing number of smart phones – estimated to be over 6 million – inside the country. In their efforts to stop this, The Korea Institute of Liberal Democracy wrote that North Korea employs over 7,000 agents engaged in propaganda and information warfare.³⁷ The UN Commission of Inquiry into Human Rights, reported in 2014 that: "The authorities engage in gross human rights violations so as to crack down on 'subversive' influences from abroad. These influences are symbolized by films and soap operas from the Republic of Korea and other countries, short-wave radio broadcasts and foreign mobile telephones."³⁸ History, current news, and world affairs are strictly controlled by the state and criticism of the government is considered a crime,³⁹ so too is the act of distributing or watching foreign media. The

---
35 Pak J, *Why we shouldn't rule out a woman as North Korea's next leader*, Brookings, 4 May 2020, available at https://www.brookings.edu/blog/order-from-chaos/2020/05/04/why-we-shouldnt-rule-out-a-woman-as-north-koreas-next-leader/.
36 DuMond M, *Information and Its Consequences in North Korea*, Beyond Parallel, 17 January 2017, available at https://beyondparallel.csis.org/information-and-its-consequences-in-north-korea/.
37 Kang T, *North Korea's Influence Operations, Revealed*, The Diplomat, 25 July 2018, available at https://thediplomat.com/2018/07/north-koreas-influence-operations-revealed/.
38 UN Human Rights Council, Report of the detailed findings of the commission of inquiry on human rights in the Democratic People's Republic of Korea, 7 February 2014, A/HRC/25/CRP.1, 366.
39 DuMond, *No Laughing Matter: North Koreans' Discontent and Daring Jokes*, Beyond Parallel, 2 November 2016, available at https://beyondparallel.csis.org/no-laughing-matter-north-koreans-discontent-and-daring-jokes/.

'Crime of Possessing or Bringing in Corrupt and Decadent Culture' can result in a prison sentence of between 3 and 15 years. In October 2020, the captain of a North Korean fishing boat was put to death for listening to banned broadcasts while fishing in the water off the coast of the country. Turned in by one of his crew members he was charged with "subversion against the party" and executed by firing squad, in front of 100 other fish boat captains.[40]

However, it is North Korea's external IO that are perhaps most interesting, and they have a longer history than is widely known. In the late 1990s the role of the North Korean Army in the Vietnam War began to finally seep into the public domain. Most of the information concerned the presence of North Korean pilots flying combat missions over northern Vietnam in 1968-69 however it also transpires that North Korean psychological warfare personnel worked alongside the Viet Kong between 1965 and 1972, targeting the South Korean Army contingent that was fighting alongside the US and South Vietnamese. In 2010 the official journal of the People's Army of Vietnam, discussed the presence of North Korean 'specialists' who helped to produce Korean-language propaganda broadcasts directed against South Korean troops.[41]

In recent years most of the IO activity has either been international cyber-attacks or more conventional leafletting over the border into South Korea. For example, in April 2016, North Korea sent over 20,000 leaflets and 40 compact discs criticizing then South Korean President, Park Geun-Hye.[42] In July 2016, North Korea sent plastic bags, each carrying about 20 leaflets threatening missile attacks, down the Han River which flows through Seoul. Some of the leaflets contained messages proclaiming that the North won the Korean war, a regular revisionist theme, and that the armistice was only signed because the UN forces begged North Korea to do so. In 2020 North Korea announced it was preparing to send 3,000 balloons over the border, each one carrying leaflets, cigarette butts and other assorted rubbish.[43] In

---

40 Miller J, *North Korea executed fishing captain for listening to banned radio: report*, The New York Post, 18 December 2020, available at https://nypost.com/2020/12/18/north-korea-executed-man-for-listening-to-banned-radio-report/.
41 "Nh ngày "tác chiến" trên làn song" ("Remembering the Time of "Battles" over the Airwaves"), Quân Đội Nhân Dân, 28 April 2010 accessed via https://www.wilsoncenter.org/blog-post/north-korean-psychological-warfare-operations-south-vietnam.
42 Friedman & Rouse, *North and South Korean Psychological Warfare*, available at http://www.psywarrior.com/korea.html.
43 Sang-Hung C, *North Korea Vows to Dump Millions of Leaflets and Trash on the South*, The New York Times, 22 June 2020, available at https://www.nytimes.com/2020/06/22/world/asia/north-korea-leaflets-south-korea.html.

retaliation South Korea has also utilised balloons to carry messages across the heavily fortified borders together with giant loudspeakers broadcasting anti-regime messages. Recognising that these tit-for-tat events were probably of little benefit, and certainly served to antagonise and rile the north, the South Korean National Assembly, in December 2020, approved legislation that imposed stiff fines and jail terms for sending leaflets, USB sticks, Bible verses, and even money into North Korea via balloons with South Koreans facing fines of up to $27,000 and up to three years in prison for violating the law.

The balloon escapades look amateurish, and the north has in recent years tried to improve and diversify its IO output utilising social media. An investigation by Google found that since 2019, North Korea state-backed hackers have targeted news outlets, journalists, and their related contacts to plant false stories and launch disinformation campaigns.[44] Much of North Korea's IO are now in the cyber domain. Perhaps the most (in)famous of these is North Korea's involvement in the hacking of Sony Pictures, in response to their release of a movie called *'The Interview'* which caricatures the Dear Respected Leader. The hack led to the temporary disablement of the studio's computer systems and the leaking of private information and documents online.[45] The DPRK also used cyber-attacks to steal currency. In 2021 the US charged three North Korean computer programmers with conspiring to steal $1.3 billion in money and cryptocurrency through state-sponsored cyber hacks. US officials told the media they believed that the three defendants were members of a specialist unit within the RGB which they code named the Lazarus group.[46] In the UK in May 2017, the National Health Service (NHS) computer systems were temporarily paralysed by a ransomware virus called WannaCry. It disabled computer systems and demanded money for their unlocking.[47] Although it is not thought that the NHS was deliberately targeted, more it was an unfortunate victim of 'collateral damage', it does indicate the sophistication of North Korea's

---

44 Stone J, *Google catches North Korean, Iranian hackers impersonating journalists in phishing efforts*, Cyberscoop, 26 March 2020, available at https://www.cyberscoop.com/google-phishing-warning-hacking-campaign-iran-north-korea/.
45 Gibbs S, *Did North Korea's notorious Unit 121 cyber army hack Sony Pictures?*, The Guardian, 2 December 2014, available at https://www.theguardian.com/technology/2014/dec/02/north-korea-hack-sony-pictures-brad-pitt-fury.
46 Manson K, *North Koreans charged in $1.3bn cyber-hack spree*, The Financial Times, 17 February 2020, available at https://www.ft.com/content/3556c699-8bff-4837-abb1-1ef70d2094d.
47 *NHS cyber-attack was 'launched from North Korea'*, The BBC, 16 June 2017, available at https://www.bbc.co.uk/news/technology-40297493.

cyber and is in stark contrast to the almost comedic balloon efforts that we saw earlier.

In 2020 the BBC aired a two-part documentary called 'The Mole'. It told the seemingly unlikely story of an unemployed Danish chef, Ulrich Larsen, who spent ten years infiltrating the Korean Friendship Association (KFA), rising through its ranks and being given more and more responsibilities. KFA is a Spanish based friendship association established in 2000 and today claims representatives and affiliates in over 20 nations including the US and UK. The documentary used hidden (and strangely often overt) cameras to uncover a network of affiliations and individuals actively supporting and promoting North Korean policies and leadership. As well as attempting to broker deals for (illegal) inwards investment into North Korea, its affiliated have attempted to counter the "slanderous propaganda that depicts DPRK as a dictatorship", or in other words "lobby on behalf of the regime".[48] Although their membership in the UK is small (some estimates put it at no more than a few hard core members focussed around British communist and regular commentator Dermot Hudson),[49] even a casual inspection of some of its affiliate's Facebook pages[50] shows it has a large and proactive following deploying very focussed messages. For example, "The Democratic People's Republic of Korea is a land of bliss where there is no unemployment and no taxation, and it has free housing, absolutely free education"[51]; accompanying a photograph of a Korean steel worker, "A photograph of one of the ruling class in the Democratic People's Republic of Korea, that is the working class... No billionaires and out of touch politicians in charge, but "ordinary" working people running their country."[52]

It is tempting to dismiss such messaging as the ranting of a few disturbed and misguided individuals, indeed its footprint and activities in Europe and the US are almost comical. But of course what it does do is allow the DPRK leadership to parade its western membership, who make regular

---

48 Kingstone K, *Meet North Korea's UK fan club: 'They have their own way of doing things'*, The Guardian, 22 September 2015, available at https://www.theguardian.com/world/2015/sep/22/north-korea-friends-of-korea-michael-chant.
49 Demetriadi A, *The Western friends of North Korea*, The Ferret, 4 March 2020, available at https://theferret.scot/north-korea-friends-western/.
50 *Staffordshire Branch of the UK Korean Friendship Association*, Facebook, available at https://www.facebook.com/Staffordshire-Branch-of-the-UK-Korean-Friendship-Association-426792611008845/?ref=page_internal.
51 Ibid., Facebook Entry 23 February 2021.
52 Ibid., Facebook Entry 11 February 2021.

visits to Pyongyang, to the North Korean population and 'show' how loved The Dear Leader is abroad. Where the KFA does present danger is in South Korea where they have a far larger membership and have actively tried to recruit political groups and figures from the left wing of Seoul politics. The KFA are Pyongyang's own 'useful idiots' (see Chapter 10) and are rewarded with various medals and titles that no doubt play to their personal vanity and perhaps compensate for the less than spectacular lives they lead in their own countries.

And what of US elections? Of Brexit? Did North Korea attempt to exert influence in the same way as Russia and Iran? David Maxwell says they know about it; "they track it; they understand it; but to deliberately manipulate it? No so much".[53] However, he and many other experienced analysts are far less sanguine about the future. Undoubtedly cyber-attacks will continue – not least as they are a source of much needed foreign currency – and will no doubt become more and more sophisticated. But the real fear is that North Korea, Russia, and Iran might in the future begin collaborative and coordinated IO projects against their mutual adversary – the US and its allies.

The People's Republic of China

One of the most frequently quoted military philosophers is Sun Tzu, a Chinese general, strategist and philosopher who lived during the Eastern Zhou period of 771 to 256 BCE. He is credited with writing *'The Art of War'* which sets out the idea that one should aim to achieve victory without direct fighting – the very essence of IO.

For a great many years, China was obscured to international public scrutiny; its backwards and largely agriculturally based economy and huge uneducated population fell victims to Mao Tse Tung's cultural revolution in the 1960s. It was not until the late 1970s that President Deng Xiaoping began changing the economy and accelerating economic development. The Tiananmen Square massacre in 1989 paused reforms but by the early 2000s China's economy was roaring, overtaking Japan as the world's second largest economy in 2010 and in 2017 overtaking the USA.[54] And

---

53 Interview Tatham / Maxwell Senior Fellow Foundation for Defense of Democracies, Zoom, 11 March 2021.
54 Elegant N, *China's 2020 GDP means it will overtake U.S. as world's No. 1 economy sooner than expected*, Fortune, 18 January 2021, available at https://fortune.com/2021/01/18/chinas-2020-gdp-world-no-1-economy-us/.

with economic prosperity came military investment, a growing appetite for global influence and a need to secure 'the home front'. In this latter regard the Chinese government has been incredibly successful in overwhelming their own population through the control of news to ensure the primacy and continuance of the communist party. In what is referred to as the 'great firewall', Chinese population access to the internet has been increasingly filtered. As a 2020 Human Rights Watch report said, "in recent years as a crackdown on the internet and civil society has become more thorough and sophisticated—and the government's messaging has grown more nationalistic".[55] The extent of that censoring is described by Bellingcat: "Access to non-Chinese websites including Google, Facebook and YouTube are blocked… to access the mainland China versions of many of these [Chinese] platforms you are required to sign up with a Chinese phone number and SIM. The process of obtaining a Chinese SIM involves linking it to your real identity".[56]

With the domestic population largely controlled China is now looking to apply the same ideas and techniques to find ways to overwhelm international populations.

Some years ago, I was commissioned to write a paper on IO for the US Strategic Studies Institute.[57] In that paper I wrote how the Chinese government was increasingly recognising that it needed to proactively influence global events to its advantage and as a result had developed a philosophy that it called the three warfares (*san zhong zhanfa*) concept – the mastery of psychological warfare, media warfare, and legal warfare. In 2003, the Chinese government endorsed that idea and began developing their IO capabilities. In 2012, a US report was published that opined that the Chinese were now on their way to developing advanced influence capabilities. The People's Liberation Army (PLA) had, apparently, developed capabilities referred to as 'Assassin's Mace' (*sha shou jian*), designed to give technologically inferior military advantages over technologically superior adversaries, and thus change the direction of a conflict.

---

55 Wang Y, *In China, the 'Great Firewall' Is Changing a Generation*, Human Rights Watch, 1 September 2020, available at https://www.hrw.org/news/2020/09/01/china-great-firewall-changing-generation.
56 Killing A, *The Challenges of Conducting Open Source Research on China*, 18 April 2023, Bellingcat, available at https://www.bellingcat.com/resources/2023/04/18/china-challenges-open-source-osint-social-media/.
57 Tatham S, *U.S. Governmental IO and Strategic Communications: A Discredited Tool or User Failure? Implications for Future Conflict*, available at https://press.armywarcollege.edu/monographs/508/.

The PLA Daily newspaper often has articles on Information Warfare. In August 2022 it contained an article about what future warfare might look like from the Chinese perspective. It started with the assertion that "information is the king of combat" and if China wished to triumph then it had to engage in three specific programs of work. Firstly, "China must accelerate the construction of cognitive offensive and defensive combat ... and build an internal talent base (presumably PLA Information Operators) who would provide all-round support for cognitive offensive and defensive combat". Secondly, China needed to accelerate the building of a "media communication matrix", that would break through the "barriers of information connectivity" (and it is not entirely clear from the article what this means) and thirdly, "China had to accelerate the coupling and linkage of information and cognitive domain operations, vigorously develop core technologies such as neural network systems, artificial intelligence applications, cognitive decision-making and psychological attack and defense".[58] Even if the exact nature of the work they are undertaking is unknown, the fact that China has been investing in IO capabilities is itself significant, particularly if they are combined with cyber capabilities – and it is worth noting that China's electronics' industry is projected to achieve revenue of US$375.70 billion in 2023, an annual growth rate of 19 per cent.[59] In February 2023, British media demanded an enquiry into the use of Chinese manufactured CCTV systems at British Army bases across the UK. The Daily Mail newspaper told its readers that: "CCTV cameras – which have also been banned by the US military amid concerns they could be used to send vital data back to Beijing's spies – at numerous military sites, including barracks for elite troops who guard the King".[60] The 'revelation' followed a spate of reports about Chinese information related activity, from tracking devices being found in UK Minister's official cars,[61] through to spy

---

58 *Aiming at the Future War and Fighting the Cognitive 'Five Battles'*, PLA Daily, 23 August 2022, available at http://www.81.cn/jfjbmap/content/2022-08/23/content_322554.htm.
59 *Electronics – China*, Statista, available at https://www.statista.com/outlook/dmo/ecommerce/electronics/china.
60 Charters & Tahr, *Chinese spy cameras are being trained on our Army: Investigation reveals glaring security lapses at British military bases as MPs demand urgent inquiry*, The Daily Mail, 25 February 2023, available at https://www.dailymail.co.uk/news/article-11793319/Chinese-spy-cameras-trained-British-Army-bases-security-lapses-MPs-demand-inquiry.html.
61 Holmes, R, *Hidden Chinese tracking device 'found in UK Government car' sparks national security fears*, iNews, 6 January 2023, available at https://inews.co.uk/news/hidden-chinese-tracking-device-government-car-national-security-2070152.

balloons over the US and speculation that Chinese researchers could break modern encryption with quantum computers.[62]

China of course has 'form' for collecting data surreptitiously. The social media app of the 'moment' appears to be TikTok – the short video hosting service that seems to have caught global imagination. By the end of December 2022, it was believed that TikTok – that also goes by the Chinese name Douyin – had over 1 billion users. Tik Tok, which is the international brand name – was launched into the market in 2017 and quickly became one of the most downloaded apps on both Android and iOS. Whilst enormously popular with children and Hollywood celebrities alike, Tik Tok collects more user data than any other similar app – beating even the Russian Facebook equivalent, Vk. In a trial undertaken by Australian company, Internet 2.0, TikTok scored 63.1 in their Index – most other apps scoring between 28 and 34 for data collection (the higher the score the more intrusive the app). Internet 2.0 found that Tik Tok had nine active trackers in the code – including the Russian app Vk SDK. Amongst the data that Tik Tok was collecting was location, calendar appointments, and contact details – which the app continuously asks if the user had previously disabled the option.

But it is China's expansionist ambitions that perhaps provides the most obvious signs of PsyOps type campaigns. There has long been a dispute between India and China – occasionally erupting into conflict – over the border of Eastern Ladakh, high in the Karakorum mountains. Retired Indian General, P. Rajagopa, told the Indian media in late 2020 that the Chinese were using what he referred to as "Wolf Warrior messaging". This apparently involved concerted PsyOps at both the strategic level – projecting China as the aggrieved party and stating that it was ready to again go to war with India – through to lower-level tactical type PsyOps, showcasing well-trained and acclimatised troops on the borders in an unrelenting steady stream of reports and photographs. Concurrently, surveys were released in The Global Times newspaper (a daily English language tabloid under the control of the Chinese Communist Party's flagship newspaper, the People's Daily) that stated over 90 per cent of respondents would support Beijing if it were to act militarily on India's 'provocation'.[63]

Whilst a remote mountain border is likely to be of little interest to the West, the issue of Taiwan definitely is. China of course has never recognised

---

62 McLaughlin L, *Chinese researchers are making claims that, if true, would threaten national security*, NPR, 18 January 2023, available at https://www.npr.org/2023/01/18/1149855926/chinese-researchers-are-making-claims-that-if-true-would-threaten-national-secur.
63 https://thedailyguardian.com/psychological-operations-is-the-dragon-winning/.

the independence of Taiwan and regards it as a renegade province which will in time be reunited with the mainland. Taiwan and much of the international community have a different view and the PRC has been exhaustive in pushing its claim on Taiwan through not just IO, which are extensive and have been described as a 'live information war',[64] but also through its regular military presence close to the island. In parallels with Russia's efforts with the Russian speaking diaspora in Eastern Europe, there is a view amongst China watchers that for all their efforts China has not been very good at making the case for reunification within the Taiwanese population who likely would be key to justify any military incursion and to ensure that China does not end up in their own version of Afghanistan or Ukraine.[65]

This does not mean that China has not attempted to influence internal Taiwanese politics. Leading Taiwanese research agency 'Double Think Lab' has tracked Chinese activities through the 2018, 2020 and latterly the 2022 elections. After the 2022 elections the agency published three key findings. Firstly, Chinese efforts at influence were becoming harder and harder to observe (although the problems caused by COVID may have meant there was just less activity). Secondly, in past elections they had been visibly supporting specific candidates – this time around they were focussed less on people and more on the amplification of specific issues that were favourable to China. Finally, they worked very hard to undermine US influence. Facebook proved to be their social media channel of choice closely followed by YouTube and a recurring theme across all of the various channels was that the US was abandoning Taiwan. Other themes were the 'chaotic' Taiwanese government (such as amplifying the drowning of Taiwanese citizens in the southern city of Kaohsiung) and the accidental breakage of ancient Chinese artefacts at the National Palace Museum. Whilst both of these events were trying, there has also been suggestions that Beijing has spread fake news – a particular example being the suggestion that the Japanese government had ordered the evacuation of all Japanese citizens from Taiwan. And overarching all of these is the expansion of China's media footprint – including TV channels based inside Taiwan and funded directly from Beijing.

---
64 Doshi R, *China Steps Up Its Information War in Taiwan*. Foreign Affairs, 15 January 2020, available at https://www.foreignaffairs.com/articles/china/2020-01-09/china-steps-its-information-war-taiwan.
65 Prasso, Sheridan; Ellis, Samson, *China's Information War on Taiwan Ramps Up as Election Nears*, Bloomberg, available at https://www.bloomberg.com/news/articles/2019-10-23/china-s-information-war-on-taiwan-ramps-up-as-election-nears.

The global COVID pandemic of 2020-2022 caused China significant problems and as the virus spread across the globe, awkward questions were asked of the Chinese authorities over how honest they had been about the source and scale of the pandemic. China launched a massive Information Operation across global social media channels to try to deflect blame away from Beijing. In June 2020, Twitter (now 'X') shut down 23,750 primary accounts and approximately 150,000 booster accounts which were being used by China to push what it called "deceptive narratives and spread propaganda".[66] Whilst the rest of the world began opening up again in Autumn and Winter 2022, China remained resolutely locked down with some of the severest COVID restrictions anywhere in the world. Its population appeared to have finally had enough and widespread protests broke out. In an attempt to hide the protests from the global community, Twitter (now 'X') in particular was flooded with posts attempting to hide protests in China, in what the media described as "an apparent state-directed attempt to suppress footage of the demonstrations … Chinese bot accounts are being used to flood the social networking service with adverts for sex workers, pornography and gambling when users search for a major city in the country, such as Shanghai or Beijing, using Chinese script".[67]

Away from Taiwan, China has also not been averse to interfering in domestic western politics and none more so than in the US. in August 2020 Google removed 2,500 YouTube videos which they suspected were of Chinese origin and included coverage of Black Lives Matter.[68] The death of George Floyd at the hands of the Minneapolis police and the wider Black Lives Matter (BLM) movement presented China with excellent opportunities to highlight what it sees as the fault lines in democracies. It was also an excellent opportunity to deflect criticism over Beijing's behaviours in places such as Hong Kong when peaceful protests were brutally suppressed. After the US government criticized China's actions, the Chinese foreign ministry mockingly responded: "I can't breathe".[69] And, finally, back to Ukraine. In

---

66 Conger, K, *Twitter Removes Chinese Disinformation Campaign*, The New York Times, 11 June 2020, available at https://www.nytimes.com/2020/06/11/technology/twitter-chinese-misinformation.html.
67 Milmo & Davidson, *Chinese bots flood Twitter in attempt to obscure Covid protests*, The Guardian, 28 November 2022, available at https://www.theguardian.com/technology/2022/nov/28/chinese-bots-flood-twitter-in-attempt-to-obscure-covid-protests?CMP=Share_iOSApp_Other.
68 Strong M, *Google removes 2,500 China-linked YouTube channels*, Taiwan News, 7 August 2020, available at https://www.taiwannews.com.tw/en/news/3982817.
69 Hua Chunying 华春莹 on Twitter: *"I can't breathe"* available at https://t.co/UXHgXMT0lk" / Twitter.

February 2023, the Head of the US State Department's Global Engagement Centre told an audience in London that China has been spending billions to spread disinformation and that most of the messaging is completely in alignment with Russia on Ukraine.[70]

Most of China's IO efforts are probably unknown to the wider public. That is not true of TikTok, which in 2021 surpassed 1 billion monthly active users. TikTok (the name given to the video sharing app for global audiences), or Douyin (as it is known domestically in China) has caused controversy for sometime. A 2023 report by Amnesty International criticised the app's content recommendation process and its 'invasive data collection' which is stated was a 'danger to young users of the platform by amplifying depressive and suicidal content that risk worsening existing mental health challenges'.[71] However it is the apps 'excessive data harvesting'[72] that has caused the most concern and prompted a 2020 US executive order that stated that 'TikTok's data collection could potentially allow China to "track the locations of federal employees and contractors, build dossiers of personal information for blackmail, and conduct corporate espionage".'[73]

Why the concern? TikTok is no different to most other Social Media apps in that regard. But what is different is that TikTok is owned by a Beijing based company, ByteDance, who are subject to article 7 of China's National Intelligence law, passed in 2017, which requires all citizens to 'support, assist and co-operate" with the country's intelligence efforts.[74] As a consequence, in March 2024 the US House of Representatives passed a bill requiring ByteDance to sell the social media platform to a company not based in China. If it did not, app stores including the Apple App Store and Google Play would be legally barred from hosting TikTok.[75] ByteDance denies, rigorously, that its US data is passed to the Chinese government; whilst we cannot know for certain, given the Chinese government's growing appetite for IO their denials seem highly unlikely and whilst I have

---

70 Singh A, *China spending billions on spreading disinformation about Ukraine war*, says US special envoy, World News, 1 March 2023, available at https://www.wionews.com/world/china-spending-billions-on-spreading-disinformation-about-ukraine-war-says-us-special-envoy-567178.
71 https://www.amnesty.org/en/latest/news/2023/11/tiktok-risks-pushing-children-towards-harmful-content/.
72 Tidy J, BBC News, *Is TikTok really a danger to the West?*, 19 Mar 2024, available at https://www.bbc.co.uk/news/technology-64797355.
73 Ibid.
74 Ibid.
75 Paul K, *The House passed a TikTok bill. Will the US really ban the app?*, The Guardian, 13 Mar 2024, available at https://www.theguardian.com/technology/2024/mar/13/will-us-ban-tiktok.

a few Social Media apps, TikTok is not one that I would ever contemplate downloading.

Like Iran and North Korea, China has been slow to join the IO 'club' but is now making up for lost time and significantly outpacing both the Hermit Kingdom and Iran in its appetite for IO and output.

9 It's All About Psychology

What is the connection between a music festival in Washington State, USA, and a 1973 performance by German comedian Jonny Buchardt? The connection, admittedly perhaps slightly tenuous, is that both, unwittingly, provide important insights into some of the science behind understanding human behaviours that should form the backbone of IO. Let's start with the music festival.

The city of George, in Washington State, used to be the home of the Sasquatch Music Festival. Every Memorial Day weekend indie rock bands, hip hop artists and comedians would gather in the US' far northwest corner to entertain audience drawn from across the US. The 2009 festival was headlined by the Kings of Leon rock band but in the context of IO is likely more memorable for a particular YouTube video which features not one of the bands but an exuberant lone dancer, clearly enjoying the music whilst a number of people are sat passively watching close by. The lone dancer looks slightly inebriated and is no doubt the subject of some humour amongst the rest of the crowd. However, the dancer's enjoyment and physical exuberance intensifies and as it does, so he is slowly joined by more and more people until those sat passively become the minority. Indeed, toward the video's end, people can be seen running to join the now very over-excited crowd. The video has been used to illustrate a number of different points about the human condition,[1] but in this context it shows the power of attraction and how groups can very quickly form from, seemingly, nothing. It shows the power of influence – although we do not necessarily know what triggers that group's formation, nor do we know the strength of conformity within that group, but we can visibly see people being influenced to change their behaviour from sedentary amusement to active participation. These should be issues of interest to IO practitioners because in a different context, for example during the insurgency in Iraq, we might ask the same questions of why people join violent groups or why they might make and lay improvised explosive devices (IEDs).

---

1 For example see Derek Siviers TED talk on leadership Available at https://www.ted.com/talks/derek_sivers_how_to_start_a_movement?language=en#t-16531.

So much for Sasquatch, what about our German comedian? Jonny Buchardt (real name *Herbert Günther Schlichting*) would probably be all but unknown outside of Germany was it not for the YouTube Video '*Fastnet1973 Sieg Heil*'.[2] In 1973 he performed at the Cologne Carnival to an audience of largely elderly Germans. Buchardt shouted a German sporting rallying cry[3] at them: "*zicke zacke zicke zacke*" and the audience immediately replied, in unison, "*hoi hoi hoi*". He repeated it, "*zicke zacke zicke zacke*" and again the audience shouted, "*hoi hoi hoi*". He quickly followed by shouting "*hip hip*" to which the audience responded "*hurra*'. And then, instantaneously he shouted "*Seig*" and the audience, without missing a beat, shouted back "*Heil*".

"*Sieg Heil!*" (Hail victory!) was adopted in the 1930s by the Nazi Party to signal obedience to Adolf Hitler. Under the Nazis the salute was mandatory for civilians but mostly optional for military personnel, who were allowed to use conventional military salutes until the failed assassination attempt on Hitler on 20 July 1944. Today the use of the '*Sieg Heil*' salute is illegal in both Germany and Austria. Visibly shocked at what he had done Buchardt can be seen to put his hand over his mouth before telling the crowd: "That can't be true, man! What? So many old comrades here tonight?! " The camera pans to the audiences, some who are laughing but a number look deeply uncomfortable. In the comments beneath the video one person writes: "this man brought hundreds of elderly people's inner Nazi out with just a simple hip hip hurray".

Whilst the video may be shocking today some 75 years after the end of World War II, in 1973 the war had been over just 28 years and anyone attending the festival aged over 35 would have been brought up in or served the National Socialist system and by implication have been influenced by its very effective propaganda. And the reason why Herr Buchardt's video is important to IO practitioners is that it too provides a visible example of the power of influence – in this case how long the effects of concerted and intense state-based influence can last and suddenly and unexpectedly reappear – in this example – nearly 30 years later. And, worryingly, it shows how these behavioural responses can persist despite the best efforts of, in this example, the victorious allies, to banish the Nazis and their ideology to the history books.

---

[2] Fasnet 1973 Sieg Heil, YouTube channel, available at https://www.youtube.com/watch?v=46QYGsf9IGs.
[3] Analogous to the British rallying cry "Oggi oggi oggi" (to which the reply is "Oi Oi Oi" often heard at rugby matches.

At the February 1945 Yalta Conference, US President, Franklin Roosevelt, British Prime Minister, Winston Churchill, and Soviet Premiere, Joseph Stalin, had proclaimed their desire to wipe out the Nazi party, its institutions, organizations, laws and cultural influences from the German public and its cultural life once they secured the surrender of Germany. This was later codified in the Potsdam Declaration, which stated that "all members of the Nazi party who have been more than nominal participants … are to be removed from public or semi-public office and from positions of responsibility in important private undertakings". The problem they faced, however, was that in 1945 there were eight million Nazi party members, more than 10 per cent of the population, and amongst the most educated and important professions, such as teachers, lawyers and civil servants, the percentage was far higher. In education the task was immense. Quite aside from the huge number of German schools that had been destroyed during the war, textbooks approved by Nazis were completely unacceptable for use in the post-war era, so too the syllabus and by implication a huge number of teachers. The intent was that Nazi influence in every aspect of society had to be eliminated but typically this depended upon which zone of occupation Germans lived within. Germany was occupied by the United States, the Soviet Union, Britain, and France and all took different approaches to the process, but it had some notable successes. Those who were deemed unable to be rehabilitated were prevented from being appointed to specific roles or granted specific licenses to operate (for example, to publish newspapers). Yet the changing relationship between the former allies began to have an effect and as Churchill began warning of the descent of an iron curtain[4] across Europe, so Germany began to be seen as a potential ally by the west, rather than an adversary, and the denazification programme began to wane.

Could it have been better, wider, more robust? Almost certainly. But this is not to take away from the fact that across an entire, conquered, nation a massive educational programme (which we might refer to today as de-radicalisation) was undertaken. One of the most visible components of that was the Nuremburg War Crimes trials where 22 individuals and various organizations were indicted before an International Military tribunal on four counts: conspiracy, crimes against peace, war crimes, and

---
4 Speech by Winston Churchill, Westminster College, Fulton Missouri, 5 March 1946 available at https://winstonchurchill.org/resources/speeches/1946-1963-elder-statesman/the-sinews-of-peace/.

crimes against humanity. Subsequently, the United States held 12 additional trials in Nuremberg of lower-level officials of the Nazi government, the military, the infamous Schutzstaffel (SS) as well as medical professionals and leading industrialists.

Attitudes towards Nazism were, as it was intended, forcibly changed. But, if Buchardt's YouTube video tells us anything it is that the behaviours associated with Nazisim – in particular the salute – lay dormant but not extinct and as a further comment on the YouTube page notes: "their instincts kicked in at the worst possible moment". Both these videos should compel us to understand the overlooked component of IO – psychology.

Attitudes and Behaviours

In the months after 9/11, the US asked itself very serious questions – not least of which was 'why do they hate us?' On the 28th of October 2001, former US diplomat Richard Holbrooke, in a piece for the Washington Post, asked "how can a man in a cave out communicate the world's foremost communications society?"[5] He referred to the way that the twin tower attack on New York City had appeared to have attracted so much support in elements of the Muslim world. His comment, perhaps inadvertently, was the starting gun for some of the world's biggest communications companies to begin work. It fostered a belief that techniques from marketing and advertising could be used to dissuade people from committing violent or terrorist acts. Over the next few years, the US DOD spent hundreds of millions of dollars on companies such as Leonie Industries, Bell Pottinger and The Rendon Corporation in their global messaging campaigns. In 2007 they were spurred on by the hugely respected Rand Corporation which published its *'Enlisting Madison Avenue'* paper in which it explored ways that techniques from the heart of the US advertising industry, Madison Avenue, could assist the US' global war on terror and make the world – particularly the Muslim parts of it – love the US more. The paper told its (military) readers that: "Business marketing practices provide a useful framework for improving U.S. military efforts to shape the attitudes and

---

5 Holbrooke R, *Get the Message Out*, The Washington Post, 28 October 2001, available at https://www.washingtonpost.com/archive/opinions/2001/10/28/get-the-message-out/b298b3c9-45b8-45e2-9ec7-20503dd38802/.

behaviours of local populations".⁶ As The Atlantic magazine described in its review of the Rand paper, "the authors suggest that the military could create more support for its operations, and thereby achieve greater success in conflicts like those in Iraq and Afghanistan, if it updated its 'brand' identity using models such as Apple, Lexus, and Starbucks".⁷

One of the reasons such ideas caught on with military commanders is that we in the global West live in a marketing and advertising saturated environment. In the 1970s it was estimated that the average person saw around 500-1500 advertisements a day. In 2021 it is estimated that the average person sees between 6000-10000 adverts a day.⁸ Adverts are entirely normal, and we accept that with so many they must have some effect on audiences. Because if they did not, Chanel would not have paid actress Nicole Kidman $33 million for appearing in a two minute commercial for their perfumes,⁹ Pepsi would not have spent $8.1 million on its 2001 Superbowl advert¹⁰ and the Jaguar car company would not have spent $8 million on its 2014 *'British Villains'* advert.¹¹ By the time I took command of the UK's Psychological Operations Group in 2010, the use of attitudinal (advertising styled) IO was hard-wired in to the DNA of coalition operations in Afghanistan and Iraq. What I struggled with was that behaviours, which seemed to me to be far more important than attitudes, and the science of Psychology which helped explain behaviour, seemed to have been displaced by the creativity of advertising and marketing agencies. What's wrong with that?

As it transpires, lots. In 2009, Dr David Myers, a professor of psychology at Hope College in Michigan, wrote in the 9th edition of the textbook *'Psychology'* that: "The original thesis that attitudes determine actions was countered in the 1960s by the antithesis that attitudes determine virtually nothing."¹² Or more simply, if we wish to change behaviour (which in conflict we invariably will always want) do we really need to understand

---

6 *Enlisting Madison Avenue The Marketing Approach to Earning Popular Support in Theatres of Operation*, The Rand Corporation, available at https://www.rand.org/pubs/monographs/MG607.html.
7 Review of Enlisting Madison Avenue by The Atlantic magazine January / February 2008.
8 *How Many Ads Do We See A Day In 2023?*, Lunio, available at https://ppcprotect.com/how-many-ads-do-we-see-a-day/.
9 Chanel No.5 The Film – 3 Minutes Version, YouTube, available at https://www.youtube.com/watch?v=nfoMbir_Qd4.
10 Britney Spears – 'Joy Of Pepsi' Commercial – HD 1080p, YouTube, available at https://www.youtube.com/watch?v=5fugLhNbwoY.
11 British Villains – 2015 One Show Automobile Advertising of the Year Finalist, YouTube, available at https://www.youtube.com/watch?v=e7gR7EYjcP8.
12 Myers D, *Psychology*, W.H.Freeman & Co Ltd; 9th Revised edition, 28 Feb. 2009.

behaviour? The counter-intuitive nature of Myer's words remains deeply problematic for many, not least senior political and military leaders. Again and again, we see senior officers directing Information Activity that will 'alter perceptions' on the misplaced assumption this will result in an immediate corresponding change in behaviour. Such a presumption is naïve and yet over 20 years of IO experience has shown this to be a firmly held view and so it is worth spending some time looking at different case studies to try and unravel the complexity (and confusion) behind it.

In the summer of 2020, following the death in the US of George Floyd, the debate over institutional racism and bias in US law enforcement brought thousands of protestors to the streets. In 2018 the New York Police Department (NYPD) had instigated 'implicit bias' training amongst its officers. Talking to the media, the NYPD explained that the training usually consisted of "a seminar in the psychological theory behind unconscious stereotypes which can lead people to make dangerous snap judgments. For instance, unconscious associations of African Americans with crime might make cops quicker to see them as suspects".[13] In the same article Robert E. Worden, director of the John F. Finn Institute for Public Safety in Albany, N.Y, explained that: "We could certainly say that the training can be credited with elevating officers' comprehension of what implicit bias is".[14] But, and it's a big but, the researchers examined data about NYPD officers' actions on the job before and after the training. Specifically, they looked at a breakdown of the ethnic disparities among the people who were arrested and had other kinds of interactions with those officers. And in those numbers, they found no meaningful change. "It's fair to say that we could not detect effects of the training on officer's enforcement behaviours", says Worden.[15] Or in other words there is no guarantee that changing attitudes or perceptions will affect behaviours.

One of the underlying assumptions about the link between attitudes and behaviour is that we usually expect the behaviour of a person to be consistent with that person's attitudes. In psychology this is called the principle of consistency and it assumes that people are rational, always behave rationally and that a person's behaviour should be consistent with their attitude(s). There have been several experiments through the years to

---

13 Kaste M, NYPD Study: *Implicit Bias Training Changes Minds, Not Necessarily Behaviour*, National Public Radio, 10 September 2020, available at https://www.npr.org/2020/09/10/909380525/nypd-study-implicit-bias-training-changes-minds-not-necessarily-behavior .
14 Ibid.
15 Ibid.

show that is a dangerous assumption. In 1934 Richard LaPierre, a professor of sociology at Stanford University, undertook an experiment where he and a Chinese family stayed at, or visited, over 250 hotels and restaurants in the US and then, many months later, wrote to each enquiring if they would accept Chinese guests. Over 90 per cent said they would not – yet in reality only 1 of the 250 hoteliers had actually refused at the time.

LaPierre wrote up his findings in December 1934 in The Journal of Social Forces.[16] "To the hotel or restaurant a questionnaire was mailed with an accompanying letter purporting to be a special and personal plea for response. The questionnaires all asked the same question, "Will you accept members of the Chinese race as guests in your establishment"". In response 91 per cent of the latter replied "No". On the basis of the above data it would appear, LaPierre wrote, "foolhardy for a Chinese to attempt to travel in the United States".

One study does not necessarily prove anything but since LaPiere's work, this subject has been a rich vein of social science research and there have been some notable milestones along the way. For example, in 1969 the US psychologist Professor, Allan Wicker, from the University of Wisconsin, published his paper *'Attitudes versus actions: The relationship of verbal and overt behavioural responses to attitude objects'*,[17] a review of 42 previous experimental studies that assessed attitudes and related behaviours. Wicker found few studies where the correlation between attitudes and behaviours was particularly high leading him to conclude that: "taken as a whole, these studies suggest that it is considerably more likely that attitudes will be unrelated or only slightly related to overt behaviours than that attitudes will be closely related to actions".[18] One of the interesting effects of Wicker's work was that the psychology community began to realise that whilst attitudes were actually poor predicators of behaviour, behaviours were actually good indicators of attitudes – a subtle but important difference. Eight years later two of the most important researchers in the field, Azjen and Fishbein, published *'Attitude-Behaviour Relations'*[19,20] in the US

---

16 LaPeier R, *Attitudes vs. Actions*, Social Forces, Volume 13, Issue 2, December 1934, Pages 230–237.
17 Wicker A W. *Attitudes versus actions: the relationship of verbal and overt behavioral responses to attitude objects*. J. Soc. Issues 25:41-78, 1969. [University of Wisconsin, Milwaukee, WI].
18 Ibid.
19 Psychological Bulletin 1977, Vol 84, No 5 888-918.
20 Handbook of Research for Educational Communication and Technology. Available at http://members.aect.org/edtech/ed1/34/34-04.html .

Psychological Bulletin which determined that there was no direct causal path from attitudes to subsequent behaviour.

This is not to say that attitudes play no part in behaviours. In further research in 1988,[21] Ajen wrote that: "the theory of reasoned action is based on the assumption that human beings usually behave in a sensible manner; that they take account of available information and implicitly or explicitly consider the implications of their actions... a person's intention to perform (or not perform) a behaviour is the immediate determinant of that action". Or put another way, people's attitudes may in some circumstances help predict their behaviours but it's not a simple relationship. In 2001 the US association for educational communications and technology published a handbook for research. In the chapter on attitudes and behaviours it concluded that: "variables such as motivation, intention, and personality traits are intervening forces that should be considered in the attitude-behaviours formula".

But why is this even an issue for IO practitioners? The answer is that IO is almost always focussed on behaviours – and often bad behaviours. IO may be used to deter young people from joining terrorist organisations; to deter people from making and laying IEDS; to encourage people to join the security or police services in conflict or post-conflict scenarios; to encourage insurgent forces to lay down their weapons and reintegrate into society. The list is endless, and it almost always involves stopping or encouraging or maintaining a specific behaviour. Yet the route to that behavioural intervention has invariable been via attitudinal responses. A simplistic example of this were the many relationship building items (RBIs) that were distributed in Afghanistan in the belief that they would encourage Afghans to like ISAF and GIRoA forces and if they liked them, they would be less willing to engage in violence against them. Consequently, we saw duvets, cooking pots, cuddly toys, coats and jackets, footballs, radios and other assorted items, many of them branded with the ISAF or GIRoA logos, handed out across the country. It may have worked in some cases but to presume that across such a large nation the disbursement of RBIs would have a significant impact on the willingness of Afghans to fight was a presumption too far. Afghans are pragmatic. People with very little will take whatever they are offered – be it branded duvets from ISAF or routes to market for poppy by the Taliban – but their behaviours are based on far more complex issues. William Binney is the former technical Director of

---

21 Ajzen I, *Attitudes, Personality and Behaviour*. Milton Keynes: Open University Press. 1988.

the US' National Security Agency and was instigator of the ThinThread intelligence programme, disbanded just prior to the 9/11 attacks, and which he felt had it been running might have stopped 9/11. In the 2015 documentary *'A Good American'* he explained that: "everything is human behaviour. Human behaviour is extremely patterned. How do people operate? How do they interact?"[22] Or as a US DOD report, *'Five Lessons We Should Have Learned in Afghanistan'* stated: "what deploying soldiers really need to learn is how and why Afghans do certain things". And why people do certain things is complex, but science does already tell us quite a bit – so let's start with Facts and Emotions.

'First to the truth' has been an unofficial mantra of western militaries dealings with the media for years. If we don't know something we try not to speculate; we take time to understand what happened – if it's very serious that might involve formal investigations and enquiries – and when we are sure we provide the facts. Which is the right thing to do, and which is honourable; it show's our collective integrity and maintains our professional reputation. Sadly, however it leaves us well behind the adversary in many instances for two reasons. Firstly, the adversary is often unbothered by the truth and recent years have suggested that truth and facts, are curiously less important than they once were. Professor Siva Vaidhyanathan of the University of Virginia is the author of *'How Facebook Disconnects Us and Undermines Democracy'*. He told the BBC World Service that: "Facebook is designed to amplify any content that generates strong emotion ... anything that generates strong emotions is going to get reaction on Facebook. Clicks, likes and comments and those reactions are going to fly faster and further".[23] It seems clear, today, that emotional responses carry as much and perhaps more importance to people than fact – and that is largely a function of social media. It is profoundly worrying but it in part explains the election of Donald Trump and Brexit. Both campaigns were 'fact lite / emotion heavy' and both triumphed.

Dr Tali Sharot of University College London is the author of *'The Influential Mind'*[24] and she has tried to explain why the deployment of facts can be so ineffective in countering strongly held but factually incorrect opinions, even amongst groups who are of above average intelligence. Sharot unequivocally states, "Facts and figures often fail to change beliefs

---

22 *A Good American*, 2015 Documentary. Netflix.
23 *How Facebook Disconnects Us and Undermines Democracy*. BBC World Service Tech Tent, 13 November 2020 available at https://www.bbc.co.uk/sounds/play/w3cszhpm.
24 Sahrot Dr Tali, *The Influential Mind*, Little & Brown 2017.

and behaviours because it ignores what makes us human".[25] In essence the brain clings to its position and the presentation and repetition of counter-facts only serves to deepen that polarisation as both sides become angry. This is exactly what happened in the UK during the 2016 Brexit debate which has been largely and erroneously attributed to the work of a small British tech company, Cambridge Analytica. The reality is far more complex and understanding the psychology of behaviours (to vote or not to vote) critical.

To paraphrase Sharot's work further, she shows us that changing or influencing people's minds (and more importantly their behaviours) on any specific issue is a complex issue but can be distilled down to four variables: the nature of the current belief, the strength of confidence audiences have in that existing belief, the new piece of evidence with which an actor may be trying to influence that audience and their confidence in that new evidence. She believes that the further away the new evidence is from the current belief the harder it will be to change people's behaviours.

This was the problem for the Remain proponents in the UK Brexit debate. By the time of the referendum, the time to present the evidence of the benefits of the European Union (EU) to the British population had long passed and simply pointing out the faults in the Leave campaign argument reaped little reward. Very poor strategic communication by the EU and successive UK governments have ensured that the British public has almost no understanding of the benefits of EU membership; this is in stark contrast to continental Europe where every EU development programme is very clearly badged and promoted. And so, in the UK's poorest communities such as the industrial Northwest, in Cornwall and in the East Midlands – all recipients of significant amounts of money from the European Structural and Investment Fund – the vote to leave the EU was significant.[26]

Like the attitude – behaviour debate, the emotion – fact debate is now generating significant research. We are starting to learn, for example, about the effect of emotion on decision making. Some studies have shown that emotions can be contagious.[27] Twitter (now 'X') data seems to suggest that the more emotional an issue the greater the likelihood of retweeting

---

25 Sharot Dr Tali, *Why Facts Don't Unify Us*, available at https://www.youtube.com/watch?v=qsQ7I6bBYCA.
26 Between 2014 and 2020 Cornwall received £476m of European aid; the Tees valley in the North East received £163m of aid and the east Midlands £205m. Each of these three areas voted Leave.
27 Fowler, J.H. and Christakis, N.A. *Dynamic spread of happiness in a large social network: longitudinal analysis over 20 years in the Framingham Heart Study*. BMJ. 337, 2008.

and that they have a strong relationship to rumour spreading.[28] Some reports note that positive messaging on a subject will get more retweets[29] others find no significant difference between the propagation of positive and negative tweets. Some research suggests that the emotional content of a message affects how much users on social media engage with that message. In many instances this research is highly nuanced and like the definitions of IO that we saw in an earlier chapter, the syntax can be confusing. Take for example two related words 'Opinion' and 'Emotion'. Opinions might interest us if we want to find a good local plumber or electrician but if we want to know what might be driving more meaningful behaviour, emotion is probably going to be of more interest. Applying what we know about opinions to emotions or vice versa is likely to not always be accurate.[30] These can be easily confused, but confusing and applying research from emotions to opinions, and vice versa is deeply unhelpful.

What else should an IO practitioner be looking at? For the last 12 years, I have taught IO to senior military officers at various school and academies around the world. The lectures always begin with the same slides and almost always end with the same result. The first slide is a picture I took of the men's urinals in the former headquarters of 15 (UK) PsyOps Group in Chicksands in Bedfordshire. Three ceramic urinals sit side by side. The next picture is of a soldier, back to the camera, standing at the left most urinal and I ask the gentlemen amongst the audience this question. "You are the next person in. Which urinal do you use?". Over the 12 years I have taught, perhaps, 50 courses. With each course numbering on average of 20 male students so I have asked this question of nearly 1500 men and not one has ever said that they would use the middle urinal, closest to the soldier. Without expectation every single person said they would either wait or they would use the righthand urinal.[31]

I then put up another picture. Same urinals but this time with two soldiers, backs to camera, one at the left urinal and one at the right; the middle one is unused. I now ask the same question but this time the answers are different. By far the largest group would be prepared to use

---

28 Oh, O., Agrawal, M. and Rao, H.R. *Community intelligence and Social Media services: A rumor theoretic analysis of tweets during social crises*. MIS Quarterly. 37, 2, 2013, 407–426.
29 Ferrara, E. and Yang, Z. *Quantifying the effect of sentiment on information diffusion in Social Media*. Peer J Computer Science, 2015.
30 *Emotions Trump Facts: The Role of Emotions in on Social Media: A Literature Review* Proceedings of the 51st Hawaii International Conference on System Sciences | 2018, https://scholarspace.manoa.hawaii.edu/bitstream/10125/50113/paper0226.pdf.
31 These slides are available to view at the book's website, www.IOFFC.info.

the centre urinal, a smaller number say they would instead wait. Finally, I put up the same image of the urinals but this time I have superimposed a ceramic partition between each one. I ask the question again: "if you come in and the left one is taken which one will you use?". This time the vote is normally split between those who still prefer to use the right-hand urinal and those who now believe it is 'ok' to use the centre one. To conclude, I ask the class two further questions: "why do you use the right-hand urinal if the left one is used but the centre one is free and why is it ok to use the centre one if a ceramic screen is placed between the two?". And the results to the first question are almost always the same; they don't know. Or rather then can offer no uniform or definitive answer on which they all agree. Some will talk of culture; some of embarrassment – either for themselves or the person already there. Nor can they offer a commonly agreed answer to the second question: the ceramic divide 'makes it ok', or its 'socially acceptable'.

So what? Well the 'so what' is that I can, with a very high degree of certainty, accurately predict the behaviour of 50 per cent of the population in a specific known context. Clearly the application of such specialist knowledge is entertaining but limited yet it serves to illustrate a key point in IO. If you know your audiences well enough and understand a little of the psychology of human nature, then predicting human behaviour in certain circumstances is perfectly possible.

At the Hebrew University of Jerusalem, in 2005, a small research team studied the results of 286 penalty kicks in top football leagues and championships world-wide.[32] The analysis of those games showed that given the probability distribution of kick direction, the optimal strategy for goalkeepers is to stay in the goal's centre. However, the study also found that goalkeepers almost always jump right or left when the penalty kick is taken – in fact the study found that they did so 97.3 per cent of the time. Why is this of any relevance to a bunch of military IO officers? The rather dispiriting answer is that it illustrates why facts can be such poor tools to deploy to achieve outcomes and why emotion is so much more important.

To have reached a penalty shoot-out both teams will have played for 90 minutes of normal time and a further 30 minutes of extra time. Both teams will have played their hearts out in front of a stadium audience of

---

32 Michael Bar-Eli and Ofer H. Azar and Ilana Ritov and Yael Keidar-Levin and Galit Schein. *Action Bias Among Elite Soccer Goalkeepers: The Case of Penalty Kicks*, Ben-Gurion University of the Negev and the Hebrew University of Jerusalem, 2005, available at http://mpra.ub.uni-muenchen.de/4477/.

perhaps one hundred thousand people and a global audience, watching TV, of hundreds of millions of people. At the end of those 120 minutes the pressure that the goalkeeper and penalty taker now face must be astronomical. In the case of the English national team, between 1990 and 2012, they had been eliminated from major international championships, on penalties, six times out of ten. England manager, Roy Hodgson, told the media after England had lost their sixth penalty shoot-out, in 2012, "You can't reproduce the tired legs. You can't reproduce the pressure. You can't reproduce the nervous tension".[33]

For the goalkeeper there are no more important people on the field than his ten other teammates who will be standing, nervously, at the other end of the field, hoping the penalty will be saved. Whilst standing still might statistically yield a better result (for example, a save), if the goal were to be scored, the goalkeeper would have a difficult conversation with his teammates. Whilst he could claim that, statistically, standing still was the better decision, to his teammates it looked like he had done very little (just the opposite to the drama and energy of a left or right jump) whilst a goal scored despite the dramatic expenditure of effort would be easier to explain to sceptical and tired teammates. The Israeli scientists described it using two terms from psychology; the first was a 'bias for action' and the second was 'norm theory'. For IO its yet another example of why understanding the context and composition of the groups that people are part of is very important if we are to positively influence their behaviour. One of the principal ways we have of doing that is through understanding Locus of Control.

Locus of Control (LOC) refers to an individual's perception of where control over events that take place in their lives resides, or (in other words), who or what is responsible for what happens to them. Control can either be internal or external. Groups and individuals with an internal (or high) LOC feel that the circumstances are under their control, while groups and individuals with an external (or low) LOC feel their fate is attributable to luck or determined by people with authority. It is useful to know whether a group has an internal LOC or external LOC because there are important differences between how the two groups are motivated, how they respond to external messages and how they respond to external influences. These

---

33 Jensen J, WH Observer, 18 November 2020, *How meticulous England ended their penalty shootout pain,* available at https://whobserver.com/how-meticulous-england-ended-their-penalty-shootout-pain/.

differences have direct relevance to the degree to which behaviour can be changed and how one should go about doing so.

Internal individuals are more able to resist coercion, as a result of inward directedness. External individuals, on the other hand, may find it harder to resist coercion because they are outward oriented. Internals are better at tolerating ambiguity; tend to be less anxious, and are less prone to depression, although they are also more guilt-prone than Externals. Externals are not willing to take risks, less likely to work on self-improvement and less likely to try to better themselves through remedial work. Internals derive greater benefits from social supports.

Internal groups are more likely to work toward objectives, to plan more long term and to wait longer for an outcome of their goals. Internal groups will display a stronger sense of identity and higher group cohesion. Internal groups are more able to resist infiltration and coercion. Internal groups are more motivated. External groups are less likely to take action, especially risky action, and self-improvement is therefore less likely. External groups are less likely to become socially mobile as a whole; rather, individuals leave the group in individual mobility. Internal groups derive greater amount of resources from state and social structures.

All of this sounds very technical, but understanding these dynamics is vital if we have any hope of our influence being effective to Groups with High, or Internal LOC that feel responsible for their own fate and so messages to them need to focus on logic, cost-benefit analysis (on a group level) and demonstrate what is best for the group. Timeframes can be short or longer term. It is important to emphasize outcomes and reinforcement or future rewards if the desired action is taken. Groups with Low, or External LOC, respond to Authoritarian and directive interventions. Timeframes should be short – focus on the present and what happens now. It may also be useful to show what might happen if the desired outcome is not taken or possible undesirable outcomes.

The last hundred or so years has seen significant research into the science of behaviour. It is well within our reach to be able to apply that science to the military in order to mitigate, reduce, or even stop violent behaviour. But like any scientific discipline new challenges will always arise and one of the most interesting, and disturbing, of recent years has been that of conspiracy theories. In this next chapter we look at how, why, and what conspiracy theories are and how IO must deal with them.

10 Conspiracy Theories and Useful Idiots

A friend of mine is a successful company director and city businessman. Together with our wives we meet two or three times a year for dinner, invariably spending most of the evening discussing our shared passion for Airedale Terrier dogs. In 2019 we sat outside a pizza restaurant watching a glorious sunset followed by a beautiful moon rise. Out of the blue my friend told me he did not believe that man had landed on the moon. I was so shocked I struggled with a reply; my awkward silence quickly filled by another friend who had also joined us for dinner, who rapidly agreed and also said that 9/11 was a hoax. Having worked in the Arab world for many years, I had regularly encountered conspiracy theories – the region always seemed particularly susceptible to them especially if they involved the US – but to have two friends tell me in the garden of a Hampshire Pizzeria that man had not landed on the moon and that the events of 11 September 2001, that had defined the bulk of my military career, were both hoaxes was a shock.

Yet belief in conspiracy theories is surprisingly very widespread, indeed some studies suggest that over half the US population, for example, believe in at least one conspiracy theory[1]. Indeed concern about conspiracy theories across the US prompted the government to award a National Science Foundation Grant for a study into their prevalence which was published by the US National Library of Medicine in 2022.[2] Amongst its many findings was an implicit conclusion, belief in conspiracies is not the preserve of the mentally ill, although clearly that would be a vulnerable group of people, but the product of normal human psychology. For example, earlier we looked at the effects of bias; one that we did not explore is Proportionality Bias. This is the idea that big events must have

---
1 Orth T, *Which conspiracy theory do Americans believe?* YouGov.US available at https://today.yougov.com/politics/articles/48113-which-conspiracy-theories-do-americans-believe.
2 PlosOne, *Have beliefs in conspiracy theories increased over time*? Published 20 Jul 2022 available at https://www.ncbi.nlm.nih.gov/pmc/articles/PMC9299316/.

major causes.³ US Psychologist Rob Brotherton, wrote *'Suspicious Minds: Why We Believe Conspiracy Theories'*⁴ and examined the theories behind the shooting of US President's John F Kennedy and Ronald Reagan. Kennedy's assassination had tremendous global consequences and as a result required a proportionally large explanation. It was impossible for people to accept that this might just have been a single criminal act. Given its enormity the assassination had to be the result of a massive national conspiracy... didn't it? Yet the attempt on Reagan's life had few consequences (Reagan went to hospital, received treatment, recovered, and went on to govern for a further eight years). With few downstream consequences, it required little explanation over and above the actions of one single insane would-be assassin. "When something big happens, we tend to assume that something big must have caused it ... This is the proportionality bias".⁵ By extension therefore it is not unreasonable to think that big events (such as 9/11, the 2020 US Presidential election and Brexit, to name a few that have become synonymous with recent conspiracy theories) will cause people to search for 'big' explanations.

In philosophy there is an expression 'Occam's Razor'. Sometimes also referred to as the law of economy, it says that given two competing theories, the simpler explanation should always be preferred. It's attributed to William of Ockham, an English monk, theologian and philosopher who lived in the early years of the fourteenth-century and who wrote about 'shaving away' unnecessary complexity from problems. We will see later the use of the Keep It Simple Stupid (KISS) idea by defence manufacturers in the 1960s which we might argue followed this guiding principle of cutting away needless extraneous detail and complexity. If you follow the principle of Occam's Razor, 9/11 was indeed solely the result of terrorists flying planes into the World Trade Center; Donald Trump didn't get enough votes to be president in 2020 and Brexit was 'the free will' of the majority of the British electorate.

Unfortunately, such simplicity is rarely found in conspiracy theories where extraneous details can and do take on significant importance. A regular sparring partner to William of Ockham – at least in philosophical

---

3 Leman, P.J.; Cinnirella, M, *A major event has a major cause: Evidence for the role of heuristics in reasoning about conspiracy theories.* Social Psychological Review, Vol. 9, No. 2, 2007, 18-28.
4 Brotherton R, *Suspicious Minds: Why We Believe Conspiracy Theories* Paperback – January 3, 2017.
5 *Account for Proportionality Bias: Big Events Must Have Big Causes*, JD Supra, 27 July 2020, available at https://www.jdsupra.com/legalnews/account-for-proportionality-bias-big-63232/.

discussions – was Walter Chatton, who argued strongly against Occam's razor and instead believed that if an explanation does not easily and adequately satisfy any event, other explanations have to be considered. And so if you follow Walter Chatton's ideas, 9/11 was quite possibly masterminded by the US deep state, a massive voting fraud may well have unfairly disenfranchised Donald Trump from the US presidency and Brexit could well have been engineered by Hedge Fund managers and far right politicians for personal gain.

The inconvenient truth is that both philosophical outlooks are entirely reasonable and conspiracy theories are of themselves not necessarily unreasonable propositions. The extent to which they are carried or pursued however is a different matter. It is also perfectly possible for some people to be comfortable with a relatively simplistic explanation for one event, say 9/11 (AQ hijacked planes and flew them into the buildings), and yet be utterly convinced that another event, such as Brexit, has multiple levels of conspiracy associated with it. It is also possible for people to hold competing views about the same incident, and this leads to the condition called 'cognitive dissonance' that we discussed earlier. We saw this with the Capitol building riots when the rioters were applauded by Trump supporters and right-wing groups, who simultaneously stated that the rioters were in fact part of Antifa and not the right-wing[6].

There also tends to be a belief that it is the less well-educated sectors of society who are most susceptible to conspiracy. There is certainly some evidence connecting educational attainment to susceptibility to conspiracy theory[7] but there is also plenty of evidence to show that well-educated people are also attracted to them if a conspiracy theory fits within their pre-existing view of the world. For example, Consultant Surgeon Mohamed Iqbal Adil, who worked in the UK's National Health Service for over 30 years, was suspended in 2020 for claiming that coronavirus had been orchestrated by the elite to control the world.[8] In the US, Richard Gage, a highly qualified architect, founded the Architects and Engineers for 9/11

---

6 Rissman K, *Trump falsely claims FBI and Antifa were 'leading the charge' in Capitol attack*, The Independent 6 January 2024, available at https://www.independent.co.uk/news/world/americas/us-politics/trump-jan-6-capiotl-riot-antifa-b2474292.html.
7 Van Prooijen, *Why Education Predicts Decreased Belief in Conspiracy Theories*, Applied Cognitive Psychology, November 201,6 available at https://onlinelibrary.wiley.com/doi/full/10.1002/acp.3301.
8 Dyer C, *Surgeon who said covid-19 was a hoax has been suspended pending GMC investigation*, British Medical Journal, 6 July 2020, available at https://www.bmj.com/content/370/bmj.m2714.

Truth in 2006,[9] an organisation that claims to have over 3000 engineers and architects demanding a reinvestigation to the events of 9/11.

An absence of critical thinking skills, in any part of a population, is deeply problematic. Without it people simply don't know how to challenge ideas and statements, or even that they should. I saw an example of this in my own family. Although I was the first to go to university, my father was a particularly talented and skilled master craftsman who turned complex architectural plans into detailed scale models long before Computer Assisted Design. During the Brexit debate his social media feeds were swamped with anti-EU messages and whilst he intellectually recognised that much of it might not be right, he didn't necessarily have the critical thinking skills to know where or how to evaluate it.

One enormous complicating factor is that sometimes conspiracies do actually happen and Dr Matthew Dentith, a leading expert in conspiracy theories, believes that we should not dismiss conspiracy theories too quickly.[10] For example, in the 1950s there was a widespread belief that the US government was attempting to control the weather. This was flatly denied but in 1974 the US government admitted that they had attempted to find ways to seed clouds to make it rain over the Ho Chi Minh trail in Vietnam to impede the resupply processes of the Viet Cong. The US Secretary of State for Defense, Melvin Laid, apologised for misleading Congress.[11] More recently there has been a widespread theory that the US is spying on… well, everyone. Emails, phone calls, internet searches – all were believed to be happening and of course it was repeatedly denied by the US government. But former National Security Agency employee, Edward Snowden, revealed a stash of stolen US government documents which showed that US telecommunications company Verizon had been providing the US with virtually all of its customer's phone records. He also revealed a tool called XKeyscore that allowed the US government to read everything a user does online – globally.[12]

---

9 *9/11 is Still Killing Survivors & First Responders*, AE911 website, available at https://www.ae911truth.org.
10 *Just because it's a conspiracy doesn't mean it isn't true*, TedX talks, available at https://www.youtube.com/watch?v=zlvS-GrA00I.
11 Hersch S, *U.S. Admits Rain-Making from '67 to '72 in Indochina*, The New York Times, 19 May 1974, available at https://www.nytimes.com/1974/05/19/archives/u-s-admits-rainmaking-from-67-to-72-in-indochina-a-first-in-warfare.html.
12 Franceschi-Bicchierai L, *The 10 Biggest Revelations From Edward Snowden's Leaks*, Mashable, 5 June 2014, available at https://mashable.com/2014/06/05/edward-snowden-revelations/?europe=true.

Collectively it makes understanding conspiracy theories, and how to deal with them, very hard. There is not even a unified agreement on what is a conspiracy theory. One academic definition is the "machinations of powerful people who have also managed to conceal their role".[13] Or in other words, the world is governed by a small group of people who seek to do so for their own advantage and through anonymity. However this is a heavily contested term and indeed numerous papers have been written on trying to accurately define what a conspiracy theory even is, or is not.[14] The trouble is that the words 'conspiracy theory' provide great latitude for people to quickly dismiss or demonise dissenting views. "Politicians use it to mock and dismiss allegations against them, while philosophers and political scientists warn that it could be used as a rhetorical weapon to pathologize dissent", says psychologist Michael Wood of the UK's University of Winchester.[15]

Some research suggests that conspiracies stem from an absence not of education but of knowledge. Belief in a conspiracy stems "not from irrationality but from the fact that they have little (relevant) information and their extremist views are supported by what little they do know".[16] They go on to observe that conspiracies are additionally fuelled by group dynamics and group polarisation. The idea of the 2020 stolen US election is an excellent example of how group dynamics and polarisation – particularly when driven by social media – can carry great resonance. Here the slight nuance is that plenty of information existed about the election to show that it had not been stolen but people were either not minded, or didn't have the skills, to look for it. Even when it was presented to them because it did not fit their existing world view, or their hopes and aspirations, it was refuted. As we have already shown in the work of Dr Tali Sharot, facts can be largely irrelevant to people's decision making.

Earlier the idea of 'Locus of Control' was discussed. Individuals with a low locus of control believe that they are not in control of their lives. German psychologist Professor Roland Imhoff believes that such individuals may be particularly prone to believe in conspiracies and they fall back on internal

---

13 Bartlett & Miller, *The power of unreason conspiracy theories, extremism and counter-terrorism*, Demos, August 2010, available at http://demosuk.wpengine.com/files/Conspiracy_theories_paper.pdf?1282913891.
14 Byford J, *Towards a Definition of Conspiracy Theories*. In: Conspiracy Theories, 2011, Palgrave Macmillan, London. https://doi.org/10.1057/9780230349216_2.
15 Political Psychology, Vol. 37, No. 5, 2016, doi: 10.1111/pops.12285.
16 Sustein & Vermeule, *Conspiracy Theories: Causes and Cures*. The Journal of Political Philosophy, Vol 17, no2, 2009.

'sense making' processes to try and help them understand the seemingly inexplicable. In a sense these people become particularly vulnerable to falling for conspiracy theories in order to meet their need to 'understand' what is happening.[17] Again the 2020 US election is a prime example. Donald Trump polled more votes than any other presidential candidate in history – apart from Joe Biden, his opponent, who got even more. This is inexplicable to hardened Trump supporters, dismayed that their candidate didn't win. Trump preyed upon those insecurities and disbelief in his 'Big Lie' rhetoric, repeating again and again that he had won.

And what of conspiracists? It is easy to presume that all conspiracists believe in conspiracies, yet this may not be true. People spread, or even start conspiracy theories, for any number of reasons. The most obvious is financial gain. We met Mr Alex Jones in an earlier chapter; he of the Sandy Hook and Pizza Gate 'fame'. But what is less well known is that Mr Jones runs a very successful business marketing male vitality and brain improver supplements which are extensively promoted via his Infowars site to a largely white, male majority audience.[18] Is it possible his sensational rants are merely a tool to draw more and more customers to his website? Or does he truly believe what he says? Actually, Mr Jones may not believe in what he says, indeed he has 'admitted' that: "I basically thought everything was staged, even though I'm now learning a lot of times things aren't staged… you don't trust anything anymore".[19]

Some conspiracists seek uniqueness. Going along with the crowd is counter to their personal need for recognition, even when that recognition brings notoriety. This neediness was explored by Professor Antony Lantian who concluded that "people high in need for uniqueness should be more likely than others to endorse conspiracy beliefs because conspiracy theories represent the possession of unconventional and potentially scarce information".[20] One example may be former British footballer and sport's

---

17 Imhoff, R., Dieterle, L., & Lamberty, P. (2021). *Resolving the Puzzle of Conspiracy Worldview and Political Activism: Belief in Secret Plots Decreases Normative but Increases Nonnormative Political Engagement*. Social Psychological and Personality Science, 12(1), 71-79, available at https://doi.org/10.1177/1948550619896491.
18 Williamson E, *Conspiracy Theories Made Alex Jones Very Rich. They May Bring Him Down*, The New York Times, 7 September 2018, available at https://www.nytimes.com/2018/09/07/us/politics/alex-jones-business-infowars-conspiracy.html.
19 Associated Press, *Conspiracy theorist Alex Jones blames 'psychosis' for his Sandy Hook claims*, 30 March 2019, available at https://www.theguardian.com/us-news/2019/mar/30/alex-jones-sandy-hook-claims-psychosis.
20 Lantain A, *I know things they don't know' The role of need for uniqueness in belief in conspiracy theories*. Social Psychology Vol 48 No 3 2017.

commentator, David Icke, who claims he is the son of God and believes that the British Royal Family are lizards.[21] Like Mr Jones, Mr Icke has built a significant business out of his books and lectures.

Some conspiracists are simply mischievous individuals who probably mean no harm and are merely looking for a 'bit of a laugh'. As a saxophonist, the 'Holy Grail' of 'riffs' is Gerry Rafferty's Baker Street and I have spent many hours trying to sound like professional saxophonist Raphael Ravenscroft who recorded the opening bars that have come to define the track and led to its iconic status. Yet there exists a conspiracy theory that it was not Ravenscroft at all but instead a British TV show host called Bob Holness. The theory stems from the influential music magazine, NME, which in 1990 wrote in its 'Believe it or Not' columns that Holness and not Ravenscroft had played the saxophone.[22] Music writer, Stuart Maconie, once NME's assistant editor, told BBC News that: "My personal and silly part in a sad story is that as an NME writer I invented the urban myth claiming that Bob played the sax solo". But Tommy Boyd, a DJ on London's LBC radio also claimed that he had made it up and, bizarrely so too do Gerry Raftery himself who told the BBC that "I made that up because I used to be asked 20 or 30 times if I was the person who did it, so to a foreign journalist I said it was Bob Holness... It was just a bit of fun".[23] Who knows where the truth lies over the conspiracy's origin, but it is interesting how otherwise seemingly rational, professional, people are happy to have their name associated with the origin of a conspiracy.

Where else might we find conspiracists? Sadly one obvious answer is politics. Donald Trump is alleged to have promoted a great many conspiracy theories during his time in office – from suspicion of the death of US Supreme Court Justice, Antonin Scalia, the cheering of Muslims in New Jersey on 9/11 (which he claims he personally witnessed), the legitimacy of climate change and of course the so called 'Big Lie' that he won the 2020

---

21 *Who is David Icke? The conspiracy theorist who claims he is the son of God*, Sky TV, 3 May 2020, available at https://news.sky.com/story/who-is-david-icke-the-conspiracy-theorist-who-claims-he-is-the-son-of-god-11982406.
22 Leigh S, The Independent, *Raphael Ravenscroft: Saxophonist best known for the riff on 'Baker Street'*, 6 November 2014, available at https://www.independent.co.uk/news/obituaries/raphael-ravenscroft-saxophonist-best-known-for-the-riff-on-baker-street-9844903.html.
23 *Why do we think Bob Holness was the Baker Street saxophonist?* BBC News, 5 January 2011, available at https://www.bbc.co.uk/news/magazine-12120809.

Presidential election.[24] Trump has "weaponized motivated reasoning" says Peter Ditto, a social psychologist at the University of California.. He "incited a mob, and weaponized natural human tendencies".[25] US Congresswoman Marjorie Taylor Greene is a vocal supporter of QAnon. "There's a once-in-a-lifetime opportunity to take this global cabal of Satan-worshiping paedophiles out, and I think we have the President to do it", she said, referring to Trump.[26]

In the UK Conservative Party former minister, Desmond Swayne, encouraged anti-lockdown street protesters to 'persist' and suggested NHS capacity figures were being 'manipulated' to exaggerate the impact of the COVID-19 virus. Conservative Member of Parliament, Adam Afriyie, has shown support for the views of Ivor Cummins, an Irish chemical engineer who promotes discredited claims about the pandemic.[27] Cummins has compared public health strategies to Nazi propaganda and allegedly retweeted a photo of the entrance of Auschwitz Extermination Camp in which the infamous slogan above the gates had been doctored to read *"Vackzine macht frei"* [vaccine makes you free].[28,29]

Indeed, the COVID-19 pandemic attracted all manner of conspiracy theories, globally. In innumerable surveys undertaken throughout the pandemic quite bizarre perceptions were uncovered.[30] 60 per cent of British adults believed to some extent that the government was misleading the public about the cause of the virus; 40 per cent believe to some extent the spread of the virus is a deliberate attempt by powerful people to gain control; 20 per cent believe to some extent that the virus is a hoax.

---

24 Kranz M, *24 outlandish conspiracy theories Donald Trump has floated over the years*, Business Insider, 9 October 2019, available at https://www.businessinsider.com/donald-trump-conspiracy-theories-2016-5?r=US&IR=T#questions-about-president-obamas-birth-certificate-2.
25 Kramer J, *Why people latch on to conspiracy theories, according to science*, National Geographic, 8 January 2021, available at https://www.nationalgeographic.com/science/article/why-people-latch-on-to-conspiracy-theories-according-to-science.
26 Steck E, *The congressional candidates who have engaged with the QAnon conspiracy theory*, CNN, 30 October 2020, available at https://edition.cnn.com/interactive/2020/10/politics/qanon-cong-candidates/.
27 Quinn B, *Take action over MPs linked to Covid conspiracy figures, PM told*, The Guardian, 4 February 2021, available at https://www.theguardian.com/world/2021/feb/04/take-action-on-mps-supporting-covid-conspiracy-figures-pm-told.
28 Ibid.
29 Marsh S, *Covid: Totnes concerns reflect UK-wide rise in conspiracy theories*, The Guardian, 11 Nov 2020 available at https://www.theguardian.com/world/2020/nov/11/totnes-covid-concerns-reflect-uk-wide-rise-in-conspiracy-theories.
30 For example, see Reuters Institute, Trust in UK government and news media COVID-19 information down, concerns over misinformation from government and politicians up available at https://reutersinstitute.politics.ox.ac.uk/trust-uk-government-and-news-media-covid-19-information-down-concerns-over-misinformation

Perhaps most startling was the belief by 44 per cent that coronavirus was a bioweapon developed by China to destroy the west – more than double the 20 per cent that believed to one degree or another that Bill Gates had created the virus to reduce the world's population. Not to be outdone, one survey found that 20 per cent of respondents believed it was a weapon being spread by Muslims.

Even the former British Prime Minister, Boris Johnson, has allegedly 'dabbled' in conspiracies claiming that opposition MPs were consorting with a foreign power to frustrate Brexit.[31] This is not just confined to the right-wing of politics. In the UK the shadow education secretary, labour MP Rebecca Long Bailey, was reportedly sacked from her position after sharing an article with an anti-Semitic conspiracy theory.[32]

Internationally, a rather prolific spreader of political conspiracy theories has been post-soviet Russia which has been adept at using state media to portray everything that is bad as being the result of an anti-Russian plot conceived in the West. Even the collapse of the Soviet Union was, according to some, planned and executed by the US. However, until the invasion of Ukraine in 2022, Putin had largely kept a distance, personally, from such theories, his media seemingly needing little help. But as the New York Times noted, "since the beginning of Russia's invasion of Ukraine two months ago, the gap between conspiracy theory and state policy has closed to a vanishing point. Conspiratorial thinking has taken complete hold of the country, from top to bottom, and now seems to be the motivating force behind the Kremlin's decisions. And Mr. Putin — who previously kept his distance from conspiracy theories, leaving their circulation to state media and second-rank politicians — is their chief promoter".[33] These theories have included suggesting that the global LBGTQ+ movement is a direct threat to Russia; that the US has developed chemical weapons laboratories across Ukraine (which is now simply a vasal of the US), and, of course, that the US wants to carve up Russia into smaller parts – in particular Siberia and the Far Eastern Oblasts.[34]

---

31 Gordon T, *Johnson accused of pushing Trump-like conspiracy theory over Brexit*, The Herald, 1 October 2019, available at https://www.heraldscotland.com/news/17938234.johnson-accused-pushing-trump-like-conspiracy-theory-brexit/.
32 *Rebecca Long-Bailey sacked after sharing article containing anti-Semitic conspiracy theory*, ITV News, 25 June 2020, available at https://www.itv.com/news/2020-06-25/rebecca-long-bailey-sacked-after-sharing-article-containing-anti-semitic-conspiracy-theory.
33 Yablokov I, *The Five Conspiracy Theories That Putin Has Weaponized*, The New York Times, 25 April 2022, available at https://www.nytimes.com/2022/04/25/opinion/putin-russia-conspiracy-theories.html.
34 Ibid.

How do we make sense of this? We should perhaps consider what the former UK Primer Minister, Harold Macmillan, once described as "events dear boy, events". The 9/11 attacks are perhaps the best example of this. On September 11th 2001, around the world billions of people went about the daily business. But at 0846 hours in New York, that routine was interrupted and across the globe people stared at their TV screens in disbelief as first one, then two, then three and finally four airliners hit the ground causing varying degrees of destruction and death. Events like that are just so tumultuous that it is understandable that people cannot easily process what they have seen. Whilst the link to Al-Qaeda in Afghanistan seemed believable, the descent into war in Iraq just a few months later was far harder for people to understand and it certainly appeared that a small group of US politicians were leading the charge against the so called 'axis of evil'.[35] So incomprehensible was the decision to hit Iraq, that it is understandable that people see conspiracy in the events that caused it. The idea of 'events' is important – some research suggests that conspiratorial thinking is closely related with situations which can cause significant anxiety, for example wars, elections, and national tragedies.[36]

Dr Sinéad Lambe, a Clinical Psychologist, observed, "Conspiracy thinking is not isolated to the fringes of society and likely reflects a growing distrust in the government and institutions. Conspiracy beliefs arguably travel further and faster than ever before. Our survey indicates that people who hold such beliefs share them; social media provides a ready-made platform".[37] People use cognitive shortcuts – largely unconscious rules-of-thumb to make decisions faster – to determine what they should believe. And people experiencing anxiety or a sense of disorder, may be even more reliant on those cognitive shortcuts to make sense of the world, says Marta Marchlewska, a social and political psychologist who studies conspiracy theories at the Polish Academy of Sciences.[38] Regardless, conspiracy theories are not only a serious problem, but their ability also to influence and alarm

---

35 See the text of President Bush's 2002 State of the Union Address Jan. 29, 2002, available at https://www.washingtonpost.com/wp-srv/onpolitics/transcripts/sou012902.htm.
36 Grzesiak-Feldman, *The Effect of High-Anxiety Situations on Conspiracy Thinking*, Current Psychology vol 32,2013.
37 *Conspiracy beliefs reduce the following of government coronavirus guidance*, 22 May 2022, available at https://www.ox.ac.uk/news/2020-05-22-conspiracy-beliefs-reduces-following-government-coronavirus-guidance.
38 Kramer J, *Why people latch on to conspiracy theories, according to science*, National Geographic, 8 January 2021, available at https://www.nationalgeographic.com/science/article/why-people-latch-on-to-conspiracy-theories-according-to-science.

will continue to grow, amplified by social media. Although invariably preposterous and often totally lacking in credible evidence, they can be almost impossible to disprove. Even trying to refute can actually reinforce them because it can be dismissed as part of the conspiracy.

How to deal with it? We are highly unlikely to stop conspiracy theories – not least as they inevitably drive-up profits on many social media sites. If we cannot address the source, perhaps looking at the effect of conspiracy theories in various audiences may be the answer? To misquote Winston Churchill once again, "However beautiful the strategy, you should occasionally look at the results". The mistake that is often made is that quantity does not necessarily mean quality although sheer volume can be a persuasive metric in itself. And what do we even mean by 'effect'. Are we bothered that someone believes, passionately, that man has never walked on the moon or that LGBQT is part of an active campaign to overthrow President Putin? The answer is, probably not. We do have to become interested when actual behaviours are triggered – like the Pizzagate shooting.

Useful Idiots

The Merriam-Webster online dictionary defines a 'useful idiot' as "a naive or credulous person who can be manipulated or exploited to advance a cause or political agenda" and cites a Wall Street Journal article that notes the Russian KGB uses them to promote Russian ideas and influence.[39] We need to be careful however of labelling people simply because they have different beliefs; one of the joys of living in a democracy is the ability to freely hold and express views that may differ to the mainstream. What people do as a result of those views, particularly if they actively seek to harm or damage the very rights of democracy that they enjoy, is perhaps a clearer indicator of their allegiances.

As one senior former UK government official told me, "there are the ones who are doing their job in plain sight and the ones that operate behind closed doors. This latter group are far more dangerous, because they drip feed poisoned advice directly into the ears of senior decision-makers and policymakers. It's well known who they are, and what they do – but because of the UK's non-existent legislation against foreign agents, there is nothing that can be done because it is all perfectly legal".

---

39 https://www.merriam-webster.com/dictionary/useful%20idiot.

The invasion of Ukraine has certainly been a catalyst for alternative views. For example, the planned 'NATO NO 2 WAR' conference was cancelled by its UK venue at the last moment after the list of speakers attracted highly pejorative attention online. Journalist and film maker, Oz Katerji, claimed to be instrumental in stopping the event which he described as a "A veritable who's who of racists, atrocity revisionists, Putin apologists and literal employees of the Russian and Iranian governments". He was referring to a poster which listed the speakers. Amongst there were two former MPs, George Galloway (whose Twitter – now 'X' – account has been labelled 'Russian State Affiliated Media')[40] and Chris Williamson, former Labour MP for Derby. Both he and another speaker, Professor David Miller, who's Twitter (now 'X') account was reportedly labelled as 'Iran State Affiliated Media'[41], host or have hosted TV programmes on Iran's Press TV channel. Also attending was Irish MEP Clare Daly, who as well as voting against an EU resolution condemning the Russian invasion of Ukraine has used her political platform to state her view that the "White Helmets staged an attack on the civilian population of Douma, Syria",[42] and Irish politician Michael Wallace, who has publicly called for the abolition of NATO.[43]

Making up Katerji's declared panel of "Putin apologists" was Edinburgh University Professor, Paul McKeigue, who alongside Professor Miller is a member of the Working Group on Syria, Propaganda and Media, described by The Guardian Newspaper as: "an alliance of far-left academics and researchers who claim western journalists, NGOs and others act on behalf of the CIA and MI6 to undermine the Syrian government, including faking evidence of civilian deaths and chemical attacks".[44] McKeigue came to more mainstream public attention when his lengthy correspondence with a purported Russian spy was made public.[45] Bellingcat founder Eliot Higgins tweeted: "Absolutely disgusting behaviour from Paul McKeigue… thinking he was in contact with Russian intelligence, and passing information to

---

40 https://twitter.com/georgegalloway/status/1511729349204103170?lang=en.
41 https://twitter.com/daverich1/status/1540626193225064448?lang=en.
42 O'Leary N, *Fianna Fáil MEP claims Daly and Wallace had spread 'conspiracy theory' on Syria*, The Irish Times, 6 July 2021, available at https://www.irishtimes.com/news/politics/fianna-fail-mep-claims-daly-and-wallace-had-spread-conspiracy-theory-on-syria-1.4613457.
43 Wilson J, *Abolish Nato, says Independent MEP Mick Wallace*, The Irish Times, 22 February 2022, available at https://www.irishtimes.com/news/ireland/irish-news/abolish-nato-says-independent-mep-mick-wallace-1.4809378.
44 Beaumont P, *The UK professor, a fake Russian spy and the undercover Syria sting*, The Guardian, 28 March 2021, available at https://www.theguardian.com/law/2021/mar/28/the-uk-professor-a-fake-russian-spy-and-the-undercover-syria-sting.
45 Ibid.

them".[46] Edinburgh University (and indeed the Working Group on Syria) is also home to Professor Tim Hayward. In an investigation by the BBC into his activities, one of Hayward's students told them that "He goes from talking about global financial markets [and] poverty, into this realm of conspiracy theories about [Syrian President Bashar al] Assad and Russia".[47] Any inspection of Hayward's Twitter (now 'X') account quickly reveals his anti-COVID vaccine beliefs, but he has also been accused of amplifying Russian propaganda. Referring to Hayward's retweet of a Russian government Tweet, the director of policy at the Community Security Trust, told LBC radio that: "We have seen over the years that Russia uses its propaganda as part of its war effort to cover-up what it is doing, and when British academics amplify and endorse that completely false propaganda, ... they are helping the Russian war effort".[48]

'Information Clearing House'[49] is an aggregation site for alternative commentary and 'news'. One its regular contributors is Finian Cunningham, a masters graduate in Agricultural Chemistry and former scientific editor for the Royal Society of Chemistry. In various articles (many on Russian government channels such as RT and Sputnik) he has accused the British government of "blatantly ramping up the media war against Russia" as well as "piling on demonization of Russia as a state".[50] He decided that British soldiers were being used to "set up false-flag provocations which can be conveniently blamed on Iran and thereby provide a pretext for all-out aggression against Tehran"[51] and on RT opined that Russophobia has been deliberately manufactured to hide the decline of the European Union: "The collapsing state of Western democracies has got nothing to do with Russia. The Russophobia of blaming Russia for the demise of Western institutions is an attempt at scapegoating for the very real problems facing

---

46 @EliotHiggins, Twitter 26 March 2021 available at https://twitter.com/eliothiggins/status/1375374648431538177?lang=en-GB.
47 *Students accuse lecturer of sharing Russia war lies*, BBC News, 31 May 2022, available at https://www.bbc.co.uk/news/education-61597405.
48 Harrison J, *Tim Hayward: University of Edinburgh Professor accused of sharing Russian propaganda*, The Herald, 13 March 2022, available at https://www.heraldscotland.com/news/19988269.tim-hayward-university-edinburgh-professor-accused-sharing-russian-propaganda/.
49 http://www.informationclearinghouse.info.
50 Cunningham F, *Britain Ramps Up Media War On Russia – Oh, The Irony*, Europe Reloaded, 15 July 2019, available at https://www.europereloaded.com/britain-ramps-up-media-war-on-russia-oh-the-irony/.
51 Cunningham F, *UK Dirty Ops on Iran? Beware the false flag provocations*, Europe Reloaded, 22 May 2019, available at https://www.europereloaded.com/uk-dirty-ops-on-iran-beware-the-false-flag-provocations/.

governments and institutions like the news media".[52] Cunningham is by no means the most vociferous commentator. Information Clearing House offers a home to others such as artist and political economist, Rob Urie, who writes: "The ongoing US war against Russia has elevated American-allied Nazis [Ukrainians] to the international stage as 'freedom fighters'",[53] and American Mike Wittney writes that: "Putin is not hated because he is a "KGB thug" or a "new Hitler"; that's just public relations gibberish. He's hated because he is an obstacle to the globalists achieving their geopolitical objectives ... He has been a thorn in their side for the better part of two decades and he has thrown a wrench in their loony plan to ... rule the world for the next century".[54]

Collectively politicians of all persuasions tend to fare badly in polling about their trustworthiness but academics are different. In 2016 The Times Higher Education supplement considered how trustworthy professors were and found that historically they were the third most trusted profession behind doctors and teachers.[55] Is it therefore right to be concerned when particular academics take such extreme views in support of dictatorial regimes and against democratic institutions and governments? Particularly so when they are potentially able to exercise such influence over their young students? When does fair and constructive challenge end and useful idiocy begin? What are the boundaries, if any, of free speech? These are deep philosophical questions and way beyond the remit of this book but in the US, for example the right to free speech is not absolute and the U.S. Supreme Court has ruled that the government may sometimes place restrictions upon it.

Most academics tend to exercise their views in the public domain – through their published work and their lectures – but what happens when individuals attempt to influence surreptitiously and behind the scenes on behalf of a foreign power? In an attempt to address some of these issues, the UK is introducing The Foreign Influence Registration Scheme (FIRS) as part of the National Security Bill. Under the proposed scheme, 'agents of

---

52 Cunningham F, *Russophobia a futile bid to conceal US, European decline*, RT., 13 February 2018, available at https://www.rt.com/op-ed/418687-us-europe-demise-russophobia/.
53 Urie R, *The Americans Started the US War with Russia*, Information Clearing House, 14 December 2022, available at http://www.informationclearinghouse.info/57388.htm.
54 Witney M, *Why Do Americans Hate Putin?*, Information Clearing House, 23 November 2022, available at http://www.informationclearinghouse.info/57362.htm.
55 Kelly G, *What do the public really think about academics*? The Times Higher Education Supplement. 1 June 2016, available at https://www.timeshighereducation.com/blog/what-do-public-really-think-about-academics?cmp=1.

influence' will need to be formally registered and if they do not there will be legal consequences of up to five years imprisonment and a monetary fine.[56] This is long overdue, particularly when you consider that in the US, Congress enacted the Foreign Agents Registration Act ("FARA"), which requires "foreign agents" who seek to influence US policy or public opinion, to register with the Attorney General, in 1938. What it is highly unlikely to manage is the mass of information now freely available online, much of it directly targeted at specific audiences such as UK and US policy makers. But how influential is social media and online content? Can information be placed online that will successfully influence outcomes, and can data we scraped from social media sites that can provide tangible advantage? Enter Cambridge Analytica.

---

[56] UK Government, *National Security Bill: factsheets*, 6 June 2022, available at https://www.gov.uk/government/publications/national-security-bill-factsheets/foreign-influence-registration-scheme-firs-national-security-bill-factsheet.

11 The Cambridge Analytica 'Conspiracy'

Through my long career it has been inexplicable to me why the military's use of kinetic force – bombs and bullets, which kill and maim – often seems to carry less pejorative public and media interest than the use of IO, which seek to almost always achieve effects without killing. Indeed it is not just the media and the public; on operations I was also surprised at the nervousness of senior officers in authorising IO yet having far higher risk thresholds for kinetics. I remember having to make a hasty trip to Helmand in Afghanistan, at the request of one of my officers, to speak to senior US officers who were in some instances taking weeks, if at all, to authorise the deployment of 15 (UK) PsyOps products; this at a time when kinetic engagement had never been higher. I am not a pacifist by any means, but it was seeing the violence of conflict close up that encouraged me to look for other ways to resolve issues and triggered my long journey in the world of IO. It was why I banned my team from wearing any kind of badges that suggested anything untoward[1] and why I invited the BBC to visit 15 PsyOps in 2012.[2]

Unfortunately there is now one story, arguably more than any other, that has undeservedly captivated public interest, and subsequently soiled, public understanding of military IO – Cambridge Analytica. Odd, because Cambridge Analytica, run by old Etonian, former investment banker and already millionaire by virtue of his marriage to a Norwegian shipping heiress, Alexander Asburner Nix, had absolutely nothing to do with the Armed Forces or military IO. Cambridge Analytica was however part of the SCL Group of companies – which was also parent to a much smaller subsidiary, SCL Defence, a company specialising in advanced intelligence support to IO and with whom I worked as a contractor for nearly five years.

---

1 For example, a badge showing a skull with a bullet passing through it and the words "You have been fucked by PsyOps' was very popular in Helmand amongst US PsyOps colleagues. I banned its use by my team throughout my period of command as it was simply untrue. For a picture of the badge see: https://twitter.com/markopilkington/status/577536244982353920.
2 BBC News, Psy-ops: Tuning the Afghans into radio, October 2012, available at https://www.bbc.co.uk/news/world-south-asia-20096416.

Nix was the self-styled face of 21st Century technology companies; the champion of 'big data' and, it appeared, the possessor of the 'secret sauce' that could manipulate and persuade online. He was the social media guru who had apparently got Donald J Trump, possibly the most unlikely presidential candidate ever, elected to the Oval Office. Nix had launched a take-over of the SCL group some months before getting his first US election contract – working to help Senator Ted Cruz's election campaign.[3]

SCL Defence, which unlike Cambridge Analytica did actually work in IO, comprised just four full time staff. Nigel Oakes and his brother Alex and two psychologists who worked from a tiny office in the Royal Institution in London. Although the wider SCL Group might not have been particularly interested in the defence arm, a few western armed forces who knew what it did, were. But being European militaries, they had small budgets and large bureaucracies; contracts were very few and far between, but SCL's Defence business reputation was strong even if its balance sheet was not.

On the Defence team no one seemed to know what Cambridge Analytica did, although everyone had heard (and largely, anecdotally, dismissed as hype) what Nix said they did. That was, the analysis of social media 'big data' to win elections using an apparently 'amazing' new tool called psychographics developed by Cambridge University academic, Dr Alexander Kogan, and a methodology called OCEAN. We now know that misgivings about these tools and their efficacy existed not just in SCL Defence but inside Cambridge Analytica itself and the final report of the Information Commissioners Office, released in October 2020, reported that: "Through the ICO's analysis of internal company communications, the investigation identified there was a degree of scepticism within SCL as to the accuracy or reliability of the [Cambridge Analytica] processing being undertaken. There appeared to be concern internally about the external messaging when set against the reality of their processing".[4]

For a few years SCL Defence was a 'List X' accredited company, which necessitated a series of security protocols and measures to allow SCL to handle classified materials. By 2010 (five years before Cambridge Analytica was formed), the List X clearances had lapsed and neither Nix nor anyone

---

3 In a what was subsequently described to me as a messy and acrimonious board meeting, he had pushed out SCL's founder and previous CEO, Nigel Oakes, who was then relegated to a small role running the bit of the company no one was particularly interested in; Defence.
4 Kaminska I, *ICO's final report into Cambridge Analytica invites regulatory questions*, The Financial Times, 8 October 2020, available at https://www.ft.com/content/43962679-b1f9-4818-b569-b028a58c8cd2.

from Cambridge Analytica were security cleared and thus had no access to classified materials. The security issue was significant. In the US, the emerging SCL US Defence office was in the process of building a SCIF (or Secure Compartmented Information Facility). This was an absolute requirement for US Defence work and had to be in place before they even attempted to bid for DOD contracts. So rigid are the rules that the only staff with access had to be US citizens security cleared by the US government.

However, the security firewall between the two companies was just a minor difference when compared to working practices. Cambridge Analytica conducted the entirety of their research online, using big data scraped from the web and as we have subsequently learned, particularly from Facebook. They then devised and ran downstream influence and communication programmes to support political parties. This was fundamentally different to SCL Defence who had no involvement in politics and who had developed an advanced Target Audience Analysis (TAA) methodology (the Behavioural Dynamics methodology, or BDM) which we saw in an earlier chapter. This used qualitative and quantitative field-based research (not internet based) to understand population groups to identify the influences and influencers and how counter-messages may be made more effective – the very essence of Information and Psychological Operations. They sold this methodology in two ways; they offered training to NATO allies and partners, and they would undertake the BD research programme for governmental clients in specific countries. In tandem with the research, they provided clients the means to measure the effectiveness of their [the client's] Information Activities (IA). SCL Defence never ran any of the downstream strategic communication programmes themselves; that was always left to the client.

Nigel Oakes told me that Cambridge Analytica regarded the field research process as being clunky, old fashioned and time consuming. In their view the future was 'big data', online web scraping and algorithms. But although Cambridge Analytica was presented as being new and cutting edge, it is hard to see, today, what it did differently to any other programmatic advertising practices already used across the advertising and marketing industry.[5] Nix, ever the salesman, was all over the media talking about psychographics and his companies' incredible ability to predict and influence voter's behaviour. He spoke of the 'data points' he held on every

---

5 A more detailed explanation is provided at https://www.acuityads.com/blog/2017/12/15/what-is-programmatic-advertising/.

US citizen and how "By having hundreds and hundreds of thousands of Americans undertake this survey,[6] we are able to form a model to predict the personality of every single adult in the United States of America".[7]

Some of those marketing claims, as well as being outlandish and utterly improbable in my opinion, also presented some very real data privacy issues. Worse, an association with the military and intelligence agencies seemed to be creeping, either deliberately or by association, into Cambridge Analytica's narrative. It was one that they were not entitled to and it caused concerns not just inside SCL Defence but also with certain journalists. One of those was investigative journalist Carole Cadwalladr of the UK's left leaning newspaper, The Guardian.

Cadwalladr was investigating Brexit and writing about links not just to Russia but also to the 2016 election result in the US. She had learnt that Cambridge Analytica might have had a role in both and had penned a series of reports which had also drawn in the tiny defence part of the business. SCL Defence had a little knowledge of the story in advance. Major X,[8] an experienced UK military officer serving at the time in East Africa, was contacted by Cadwalladr and asked if he would tell her any 'interesting'[9] stories about SCL. He had replied with a quote saying he knew SCL Defence and had been impressed with their work but when the story was published his quote was not included. Instead it seemed that Cadwalladr had conflated the election and defence businesses into one and stated that "This is not just a story about social psychology and data analytics. It has to be understood in terms of a military contractor using military strategies on a civilian population".[10] It was further insinuated that SCL Defence had engaged in disinformation.

These were all serious allegations. SCL Defence had spent much of the previous two years working with governments in Eastern Europe to help them understand and combat Russian disinformation and fake news,

---

6 The five-factor model. The traits include 'extroversion', 'emotional stability', 'agreeableness', 'conscientiousness' and 'intellect/imagination'.
7 Ryssdal K, *I asked a security expert to reveal how Cambridge Analytica might target me based on my personality*, 26 March 2018, available at https://www.marketplace.org/2018/03/26/i-asked-security-expert-reveal-how-cambridge-analytica-might-target-me-based-my/.
8 To protect the identity of this still serving officer I have used the nomenclature 'X' for convenience.
9 Major X shared with me screen shots of his correspondence with Ms Cadwalladr on this issue.
10 Cadwalldar C, *The great British Brexit robbery: how our democracy was hijacked*, The Guardian, 7 May 2017, available at https://www.theguardian.com/technology/2017/may/07/the-great-british-brexit-robbery-hijacked-democracy.

but spreading disinformation was absolutely not what the [Defence] part of the company did and it was incredibly damaging.

Major X immediately sent a message to Cadwalladr asking from where she had got the reference regarding the use of disinformation. Her reply to him was that she had found it on the Wikipedia website for SCL.[11] She was right, some unknown internet 'journalist' had glibly stated that SCL was engaged in disinformation and their unreferenced quote (opinion) was now being reported as a 'fact' by The Guardian. The irony of someone writing authoritatively about disinformation (deliberate) but doing so using misinformation (a mistake) was not lost. As SCL's Lead Contractor for Defence projects (which for marketing purposes I had been badged Director of Defence Operations), I emailed Cadwalladr and asked her on what basis did she substantiate the disinformation quote? Her justification was that I had previously written about disinformation and propaganda. I pointed out that Professor Peter Neumann at King's College wrote about terrorism, but no one would ever suggest that he would engage in acts of terrorism. But the damage had been done and within a few hours of The Guardian's front page splash, my phone, email, Twitter, Facebook and LinkedIn accounts went crazy as journalists, campaigners and a great many conspiracists went mad, venting their anger at Trump, Brexit, propaganda, brain washing, mind control, Psychological Operations, the CIA, the government and life in general. And all at me, a recently retired (regular) Naval officer and serving reserve officer.

Clients and acquaintances were being deluged with Freedom of Information requests. In Norway, where I (and not SCL Defence) had just finished a project with the Norwegian Defence Research Agency on mapping Russian influence in Norwegian civil society, journalist Osman Kibar of the *Dagens Naeringsliv* newspaper, worked tirelessly to uncover imagined conspiracies and wrongdoing. Kibar concluded that Cambridge Analytica was working for the Norwegian government and was making all kinds of allegations – none of them, in my opinion, baring much resemblance to the truth but causing deep distress to hard working Norwegian civil servants and, of course, immediately shutting down the project. I received a very embarrassed email from a senior Norwegian Civil servant apologising profusely, but in an effort to end the bad publicity, they had to dispense with my consultancy. And in the US, Rolling Stone magazine carried a long piece by Bob Dreyfuss opining of deep links between SCL Defence's government

---

11 Major X, personal correspondence with author, ibid.

work, retired 'spooks and right-wing US generals', my academic papers, and Russia.[12]

The fact that I had commanded the UK's 15 Psychological Operations Group for three years played badly in the coverage. Psychological Operations, for some, conjures up images of brain washing, interrogation and white noise. In reality, it is far more mundane. As I showed in Chapter 1, PsyOps is the truthful and attributable transmission of data to target audiences to persuade them to alter their behaviour. To stop them laying roadside bombs in Afghanistan for example, or to encourage the local population to support peace efforts. But this did not fit the conspiracy narrative. When I was in command of the group, we had received hate mail and, even when we had won the Firmin Sword of Peace for our work in Helmand (a story covered extensively by the BBC),[13] our work was still seen as unpleasant, underhand, and dirty. Being the former Commanding Officer of a PsyOps unit was irresistible to journalists. As Mark Laity, a former journalist and former Head of NATO Strategic Communications liked to say, "Perception is reality".

It was simultaneously fascinating and terrifying to be inside a truly global media story, and utterly disheartening to see how years of careful work building a science led approach to IO was being destroyed almost overnight. I was utterly unprepared for the media feeding frenzy, even discussing with MoD colleagues if I needed security at my house. As the company clocked over 3000 requests a day for interviews and comment everyone connected buckled under the weight of a story that led the news agenda, globally, for weeks and which still routinely appears in social media now several years later.

Whilst the 'bad guys' of the story were Nix, SCL and Cambridge Analytica the 'good guys' began to appear in the narrative. US academic, Professor Carroll, who perfectly understandably was disturbed to find that Cambridge Analytica held personal data on him and wanted to know what it was. And a Swiss researcher called, Dr Paul-Olivier Dehaye, a former professor of mathematics at the University of Zurich, also came to

---

12 Dreyfuss B, *Cambridge Analytica's Psy-Ops Warriors*, Rolling Stones, 21 March 2018, available at https://www.rollingstone.com/politics/politics-news/cambridge-analyticas-psy-ops-warriors-204230/.
13 141/2012 – The Firmin Sword of Peace awarded to 15 (Uk) Psychological Operations Group, UK MoD, available at https://www.gov.uk/government/news/141-2012-the-firmin-sword-of-peace-awarded-to-15-uk-psychological-operations-group.

prominence; it transpired that he too had been investigating Cambridge Analytica's data usage for some time.

Alongside The Guardian team, one other name stood out – propaganda researcher Dr Emma Briant of Essex University. Paradoxically Dr Briant's knowledge of SCL had largely been facilitated by me. The leading UK academic in the area of propaganda and IO, Professor Phil Taylor of Leed's University, had sadly passed away a few years before and I felt that it was important to encourage the involvement of independent academia. Given her research interests, I had invited Dr Briant to join the editorial board of the NATO Strategic Communication Journal, which I had started and of which I was the first Editor-In-Chief, on behalf of the NATO Centre of Excellence in Latvia. During one of the board meetings, Dr Briant had shown interest in the work of SCL Defence and I had encouraged SCL Defence to engage with her to demystify and dedramatize the whole subject. I had done similar with the BBC a few years before when 15 UK PsyOps Group had won the Sword of Peace for its work in Afghanistan. The BBC had been invited to report upon the work of PsyOps.[14] Naively I had hoped that by inviting an academic to see the reality of SCL Defence's work, it might serve to curb some of the media coverage that Nix was engendering, many months before The Guardian story broke.

Globally, media, academics and social media researchers were now looking at Nix's many claims about data and behavioural science and asking questions. Questions about morality, about ethics, about influence and chiefly about data privacy. The apparent closeness and obvious campaign similarities of both the US and Brexit votes was worrying many people. Could there have been malign influences at play? Many thought that there were, and whilst much of the debate was that the Russians had swung the vote, a growing view was coalescing on the idea that it was instead a small London based technology company. Nigel Oakes was uneasy and confided in me that Nix, who apparently believed all publicity was good publicity, was perfectly happy with the coverage.

The questions grew, particularly and ironically using Nix's preferred battle ground of social media to spread and find new data and researchers. His assertions seemed highly dubious to people who understood the complexity of human behaviour and influence but driven by a rabid global press and incensed at Donald Trump's election, psychographic

---

14 *Inside the British army 'psy-ops' headquarters*, BBC News, 28 October 2012, available at https://www.bbc.co.uk/news/av/uk-20067865.

modelling took on a sinister and Orwellian reputation for manipulation and influence.

Some seven years down the line from the headlines it is possible to draw some clearer conclusions. Firstly, I am utterly unconvinced that Cambridge Analytica 'won' Trump's election for him using advanced psychographics – I have seen no evidence of any worth that these apparently 'amazing' psychological tools were indeed amazing, could predict very much about anyone and certainly not the complexity of human behaviour. There is now a growing view that targeted online marketing – which is essentially what Nix was selling – is nowhere near as successful as is claimed and whilst Nix was trumpeting his OCEAN methodology and psychographics, others were moving out of the market completely.

In 2017, Proctor & Gamble cut US$200 million from their targeted online advertising spend budget, stating it to be inefficient. In 2020, AirBnB said it was making a permanent shift away from digital marketing for similar reasons. Brand Safety Advocate and marketing activist, Nadini Jammi, was asked by the BBC's World Service why, if targeting online marketing was so inaccurate, as she claimed, was so much money still being thrown at it? She told The Business Daily programme that "no one wants to admit across the supply chain we have screwed up. Lots of people would lose their jobs if they said it".[15] But perhaps a bigger bomb shell was Georgetown academic and author of '*Sub Prime Attention Crisis*', Tim Hwang, who compared the industry to a bubble that would soon burst. In a stinging criticism of the whole industry, Hwang wrote that "Big Data is not the 'new oil' It really isn't. It's the next mortgage-backed security".[16]

Cambridge Analytica certainly worked for the Trump campaign. But far from being the orchestrating mastermind behind his victory, they seemed instead to undertake much more pedestrian tasks. Indeed their work appeared not dissimilar to the so called COBS (Consortium of Behavioural Scientists)[17] who had advised and assisted President Obama's social media campaign some eight years before (and for which he was

---

15 Interview BBC World Service Business Daily, *Is the digital ad movement overvalued*, 22 April 2021, available at https://www.bbc.co.uk/sounds/play/w3ct1j92.
16 Ibid.
17 Bennett D, *Meet the Psychologists Who Convinced You to Vote for Obama*, The Atlantic, 13 November 2012, available at https://www.theatlantic.com/politics/archive/2012/11/meet-psychologists-who-convinced-you-vote-obama/321531/.

labelled as 'blind' to the dangers of social media[18]). US broadcaster CNBC queried why Trump's use of social media was labelled 'sleazy' by the media, but Obama's was seen as innovative – to which an Obama official replied: "I ran the Obama 2008 data-driven microtargeting team... We didn't steal private Facebook profile data from voters under false pretences" but offered no further substantive difference.[19]

The US' National Public Radio reported that: "according to a source close to that operation, psychological profiling never entered the picture. Given Trump's lack of digital infrastructure, the Cambridge Analytica staffers focused on much more basic goals, like increasing online fundraising, reaching out to undecided voters, and boosting Election Day turnout. People involved in both the Cruz and Trump campaigns say they never used the data Cambridge Analytica illicitly acquired from Facebook. And both campaigns ultimately soured on both the company and its CEO. Nix repeatedly claimed credit for both Cruz and Trump's triumphs, violating long-held norms for behind-the-scenes campaign operatives".[20]

Facebook's Andrew Bosworth referred to Cambridge Analytica as "snake oil salespeople. The tools they used didn't work, and the scale they used them at wasn't meaningful. Every claim they have made about themselves is garbage. Data of the kind they had isn't that valuable to begin with and worse it degrades quickly, so much so as to be effectively useless in 12-18 months".[21] Even Brad Pascale, Donald Trump's election manager, told CBS News that [psychographics] "don't work".[22] Author of *People Vs Tech*, Jamie Bartlett, notes that:

> In my opinion Cambridge Analytica's more significant role for Trump was in building 'universes' of 'persuadable

---

18 Bogost D, *Obama Was Too Good at Social Media*, The Atlantic, 6 January 2017, available at https://www.theatlantic.com/technology/archive/2017/01/did-america-need-a-social-media-president/512405/.
19 Lovelace B, *Crisis manager questions why Trump's use of social media data is 'sleazy' and Obama's 'innovative'*, CNBC, 19 March 2018, available at https://www.cnbc.com/2018/03/19/crisis-manager-why-is-trump-use-of-social-data-bad-and-obama-good.html.
20 Etrow S, *What Did Cambridge Analytica Do During The 2016 Election?*, NPR, 20 March 2018, available at https://www.npr.org/2018/03/20/595338116/what-did-cambridge-analytica-do-during-the-2016-election.
21 Lecher C, *Facebook executive: we got Trump elected, and we shouldn't stop him in 2020*, The Verge, 7 January 2020, available at https://www.theverge.com/2020/1/7/21055348/facebook-trump-election-2020-leaked-memo-bosworth.
22 Stahl L, *Facebook "embeds," Russia and the Trump campaign's secret weapon*, CBS News, 8 October 2017, available at https://www.cbsnews.com/news/facebook-embeds-russia-and-the-trump-campaigns-secret-weapon/.

voters' (e.g. American moms worried about child-care who hadn't voted before). Creative types then designed specific ads for these universes, based on the specific things they cared about. It's almost impossible to ever know what precisely made a difference when it comes to campaigning and advertising" but, significantly, Bartlett concludes that "there are no studies about whether this works for voting behaviour … I don't think millions of minds can be manipulated by … psychographic messaging.[23]

Tim Hwang told the BBC's World Service that: "we are sold a vision of digital advertising being a hyper-targeted persuasion machine... but 40% of the data is inaccurate and even when you have accurate data its only good at finding people that would have bought the product anyway".[24] There are numerous other studies that similarly suggest negligible or no impact[25,26,27] but Hwang's comment is perhaps the most insightful.

If they were doing anything meaningful at all, Cambridge Analytica was simply intensifying attraction to Trump amongst those who were already voting for him. And as we saw in the chapter on the psychology of IO, attitudes are not good precursors to changing behaviours but if someone's behaviour is already pre-ordained (such as voting for Trump or Brexit) then their attitudes can certainly be hardened. But to suggest that they were seminal in Trump's vote? A massive step too far in my view and it is hard not to note that President Trump increased his vote share in the 2020 Presidential election from 62.9 million votes in 2016 to 74.2 million in 2020, an increase of nearly 18 per cent and the second highest vote ever recorded in a US election. Cambridge Analytica was definitely not working for Trump in 2020.

But what of Brexit? Brexit for this author – proud European and utterly determined Remainer who had marched, twice, with over a million others

---

23 Bartlett J *Psychographics. What is it? Does it work? And Cambridge Analytica use it?* https://reaction.life/psychographics-work-cambridge-analytica-use/.
24 Interview BBC World Service Business Daily, *Is the digital ad movement overvalued*, 22 April 2021, available at https://www.bbc.co.uk/sounds/play/w3ct1j92.
25 Bailey M, Hopkins D, *Unresponsive and Unpersuaded: The Unintended Consequences of a Voter Persuasion Effort* Polit Behav 38, 713–746 (2016) doi:10.1007/s11109-016-9338-8.
26 Hersh E and Schaffner B, *Targeted Campaign Appeals and the Value of Ambiguity* Journal of Politics 75, no. 2 (April 2013): 520-534.
27 Nickerson D, *Political campaigns and Big Data*, The Journal of Economic Perspectives, Vol. 28, No. 2, Spring 2014.

in London to protest Brexit – remains particularly painful. Did Cambridge Analytica have a hand in what remains, seven years after the event a hugely divisive issue for the UK. During those seven years, and post scores of conversations, discussions, interviews, and research the answer appears to me, on balance, an equivocal 'no'. Brexit was categorically not the result of military grade IO (as alleged by the various Cambridge Analytica whistle-blowers) – none of the former military staff employed by SCL undertook any work on Brexit and SCL's TAA process was never involved in any part of Cambridge Analytica's political work. And, in talking to past Cambridge Analytica directors, staff and former SCL employees, it has been impossible to find anyone who believes that Cambridge Analytica undertook any meaningful work on the Brexit Referendum (save of course the 'whistle-blowers' in their best-selling books!), although it seems equally clear that Cambridge Analytica would most certainly have done so had they been given the opportunity. This narrative is clearly compounded by the opaque relationship between Cambridge Analytica and Canadian company, Aggregate IQ. There was obviously some kind of relationship and disappointingly former SCL Chairman, Julian Wheatland, who having helped me in this book subsequently never replied to my detailed questions on this particular issue.

The chief impediment to clarity remains Cambridge Analytica's principal, Alexander Nix, a man who in secret camera footage recorded by the UK's Channel 4 news, boasted of using bribes, former spies, and Ukrainian prostitutes to entrap politicians[28] but when confronted by the footage spoke of his deep embarrassment and regret.[29]

My personal view is that Nix saw an opportunity for creative ambiguity that he was somehow instrumental in the vote's outcome. It placed him at the centre of (another) global event and allowed, again in my opinion, a degree of self-interest, perhaps even narcissism, to permeate the discourse. However, one man's fantasy cannot alone justify the popular narrative that has been so energetically spread, causing huge numbers of people to believe that he and his team of 'military grade' psychological

---

28 *Head of Cambridge Analytica filmed discussing use of bribes and sex workers to entrap politicians*, The Daily Telegraph, 20 March 2018, available at https://www.telegraph.co.uk/technology/2018/03/20/head-cambridge-analytica-filmed-discussing-use-bribes-sex-workers/.

29 Mavidiya M, *Cambridge Analytica former CEO Alexander Nix revealed after Facebook scandal questioning*, The Daily Mail, 7 June 2018, available at https://www.dailymail.co.uk/news/article-5817621/Cambridge-Analytica-former-CEO-Alexander-Nix-revealed-Facebook-scandal-questioning.html.

operations staff twisted and polluted the narrative and convinced the nation to vote for the UK's accession from the European Union.

We now know that some elements of the Brexit referendum were illegal.[30] Whether its result should have been used as the determinate of such an enormous and life changing decision remains hotly contested. It did perhaps demonstrate the complete failure of the Remain political class (and of all its supporters in civil society, the media and academia) to understand huge swathes of the British public and it certainly showed the dangers people face in putting so much personal data onto social media. It also demonstrated the apparent carelessness, maybe even ambivalence, of Facebook in handling personal data and its willingness to profit from it. It also showed how the platform carried often factually dubious advertisements unchecked. Challenged by the media and asked if disinformation campaigns could distort elections, Facebook's Mark Zuckerberg refused to commit to any changes, saying that "it's really important that people can see for themselves what politicians are saying, so they can make their own judgments...I don't think that a private company should be censoring politicians or news". Over 200 of his own employees disagreed, publishing a letter asking him to reconsider, saying that "free speech and paid speech are not the same".[31] Facebook's data management and privacy practices were subsequently examined by both UK and US authorities.

In the UK, the Information Commissioner issued a monetary penalty notice under section 55A of the Data Protection Act 1998 against Facebook over its failings in the storage and handling of data. In 2019 Facebook paid the ICO £500,000, their legal counsel telling the media that: "as we have said before, we wish we had done more to investigate claims about Cambridge Analytica in 2015".[32] However, this was nothing compared to the subsequent US$5 billion fine the company had to pay to the US Federal Trade Commission (FTC) over charges that it had violated a 2012 FTC order by deceiving users about their ability to control the privacy of their personal information. That penalty is, to date, the largest ever

---

30 Graham-Harrison E, *Vote Leave broke electoral law and British democracy is shaken*, The Guardian 17 July 2018, available at https://www.theguardian.com/politics/2018/jul/17/vote-leave-broke-electoral-law-and-british-democracy-is-shaken.
31 Milman O, *Defiant Mark Zuckerberg defends Facebook policy to allow false ads*, The Guardian, 2 December 2019, available at https://www.theguardian.com/technology/2019/dec/02/mark-zuckerberg-facebook-policy-fake-ads.
32 See https://ico.org.uk/about-the-ico/media-centre/news-and-blogs/2022/02/somerset-bridge-insurance-services-statement/.

imposed on any company for violating consumers' privacy. Yet despite these fines, the company continues to be dogged with charges of data loss and mismanagement. In 2021, Facebook refused to apologise after personal data of 11 million UK users was hacked and then released online.[33] This included full names, locations, dates of birth, phone numbers and email addresses. In 2022 Ireland's data privacy regulator imposed a €265 million ($277 million) fine bringing the total it has fined Facebook's parent group, Meta, to almost 1 billion Euro[34] and in December 2022 the EU warned Meta that it could face a $11.8 billion fine (10 per cent of its annual revenue) for breaching antitrust rules.[35]

But what of Russian interference – a central allegation in the story? It is far more difficult to prove and despite formal Parliamentary investigations in the UK, only anecdotal, not empirical, evidence can be offered. Firstly, Russia might well have intervened to deliberately create chaos. Chaos is an important word in Russian political philosophy. Russian political philosopher Professor Alexander Dugin – a man who enjoys almost cult like status in some sections of Russia today and who came to global attention in August 2022 when his daughter was apparently killed by anti-Ukraine war Russians[36] – is an enthusiastic Putin supporter and has written widely on the eradication of liberalism and hegemony of the US. Dugin believes the creation of chaos an important asset in achieving that goal and the amplification of both sides of the UK's Brexit debate – both Remain and Leave – to promote chaotic UK political rhetoric and behaviour sits perfectly in that world view.

A second reason Russia may have intervened is, why not? Why would Russia not have used the many tools at its disposal to seed mayhem and distrust within a country that it has long historically hated (epitomised in the Russian phrase 'опять англичанка гадит', which (politely) translates to 'The Englishwoman is spoiling things again', a reference to the trouble

---

33 Wright M, *Facebook refuses to apologise after personal data of 11m UK users hacked*, The Daily Telegraph, 7 April 2021, available at https://www.telegraph.co.uk/news/2021/04/07/facebook-refuses-apologise-data-breach-11-million-uk-users-information/.
34 *Irish regulator fines Facebook 265 million euros over privacy breach*, Reuters, 28 November 2022, available at https://www.cnbc.com/2022/11/28/irish-regulator-fines-facebook-265-million-euros-over-privacy-breach.html.
35 Brown R, *Meta could face $11.8 billion fine as EU charges tech giant with breaching antitrust rules*, CNBC, 19 December 2022, available *at* https://www.cnbc.com/2022/12/19/meta-could-face-11point8-billion-fine-as-eu-charges-it-with-antitrust-breach.html.
36 McDonald S, *Alexander Dugin's Daughter Killed by Anti-War Russians: Former State Deputy*, Newsweek, 21 August 2022 available at https://www.newsweek.com/alexander-dugins-daughter-killed-anti-war-russians-former-state-deputy-1735497.

Queen Victoria once caused Russia. Why would Russia not intervene to disrupt the EU, an organisation that it dislikes and which it sees as an economic competitor? Why would Russia not intervene given it was already doing so in the US election, had attempted to do so in the Scottish referendum and, through its active measures programme, had done so globally for years? Why would it not do so given its written in to their own IO doctrine, in the papers from its defence academies and even its media routinely talk about it? Why would it not do so when it was clearly so easy to do? And why would it not given its apparent connections to the UK's Conservative Party?

Political activist and former major investor in Russia, Bill Browder, told Channel 4 Dispatches that: "there is a large group of British professionals, politicians, lawyers, PR Firms and private investigators who are paid by Russian oligarchs, on behalf of the Russian Government, to influence British political outcomes".[37] You do not have to search hard to find those Russian connections. Alexander Temerko, whose career took him to the top of the Russian arms industry and who it is claimed had connections at the highest levels of the Kremlin, became one of the Conservative Party's major donors – gifting them over £1 million – and close friends with Boris Johnson.[38] Evgeny Lebedev, who has dual Russian and British nationality, is the son of a former KGB spy and was awarded a Peerage by Boris Johnson (despite apparent concerns by the UK's Security Services[39]) for philanthropy and services to the media. Even the war in Ukraine has not apparently dampened donations to the Conservative Party – the media reporting tens of thousands of pounds of donations including £50,000 from Lubov Chernukhin, who is married to President Putin's former deputy finance minister.[40] Amongst the Russian analyst community – and behind the closed office doors of many UK political figures – there is a view that Russians have been able to yield influence at the top of government and they have seemingly been allowed to do so in plain sight.

---

37 *Strippers, Spies and Russian Money*. Channel 4 Dispatches documentary aired 12 February 2023.
38 Belton C, *Special Report: In British PM race, a former Russian tycoon quietly wields influence*, Saltwire, 19 July 2019, available at https://www.saltwire.com/nova-scotia/news/special-report-in-british-pm-race-a-former-russian-tycoon-quietly-wields-influence-334601/.
39 BBC, *Boris Johnson was warned of Lebedev security concerns, says Cummings*, 16 March 2022, available at https://www.bbc.co.uk/news/uk-politics-60765665.
40 Williams M, *Tories have taken £62,000 from Russia-linked donors since war began*, Open Democracy, 9 June 2022, available at https://www.opendemocracy.net/en/dark-money-investigations/conservative-party-russia-donors-ukraine-invasion/.

These issues were examined by the Intelligence and Security Committee of the UK Parliament and a report completed in 2019, was finally published in 2020.[41] Yet the report found no evidence that Russian interference had affected the Brexit referendum, not because it found no evidence but because the government had "not authorized any investigation into that matter due to the fact that any such attempt was not within the purview of British intelligence services". Or as noted Russian observer at the Chatham House think-tank, Keir Giles, commented: [the report is] "not one of Russian success but British Government failure in protecting British governance and society ... no one is in charge of protecting democratic processes in this country".[42]

However, the report did discover evidence that the 2014 Scottish independence referendum had been interfered with. This has led many academic studies to support the thesis that Russia influenced Brexit. They would point to the many open-source studies that have now been undertaken. For example, the project on Computational Propaganda run by the University of Oxford has produced a number of research papers on the issue. In their 2016 research work, Howard and Kollanyi found that automated bots played a small but strategic role in shaping Twitter conversations about the referendum. They found that the hashtags associated with the argument for leaving the EU dominated the information environment with less than one per cent of sampled accounts generating almost a third of all the messages.[43]

Specialist in social influence and networks, Oxford scientist Vyacheslav Poloski, noted that "not only were there twice as many Brexit supporters on Instagram, but they were also five times more active than remain activists".[44] Poloski goes on to say that the data analysis he undertook shows that not only did Brexit supporters have a more powerful and emotional message, they were also more effective in the use of social media... [Leave] "outmuscled its rival, with more vocal and active

---

41 *Intelligence and Security Committee of Parliament. Russia. Presented to Parliament pursuant to section 3 of the Justice and Security Act, 2013.* https://upload.wikimedia.org/wikipedia/commons/3/3d/ISC_Russia_Report.pdf.
42 Chatham Houses' January 2021 roundtable, *The Russia Report: Why Has Nothing Been Done?* https://www.chathamhouse.org/events/all/research-event/russia-report-why-has-nothing-been-done.
43 Howard, P. N., & Kollanyi, B. (2016). Bots, #StrongerIn, and #Brexit: Computational Propaganda during the UK-EU Referendum. *arXiv:1606.06356 [Physics].* Retrieved from http://arxiv.org/abs/1606.06356.
44 Polonski V, *Impact of Social Media analysis on the outcome of the EU referendum* available at http://www.referendumanalysis.uk/eu-referendum-analysis-2016/section-7-social-media/impact-of-social-media-on-the-outcome-of-the-eu-referendum/.

supporters across almost all social media platforms... led to the activation of a greater number of Leave supporters at grassroots level and enabled them to fully dominate platforms like Facebook, Twitter and Instagram, influencing swathes of undecided voters who simply didn't know what to think". To help us understand the scale of the 'battle' he concludes that "Leave outnumbered Remain by 7:1 on Twitter ... the Leave Camp was able to create the perception of a wide-ranging public support for their cause that acted like a self-fulfilling prophecy, attracting more voters to back Brexit".

But the Intelligence Select Committee concluded that: "We have not been provided with any post-referendum assessment of Russian attempts at interference. This situation is in stark contrast to the US handling of allegations of Russian interference in the 2016 presidential election, where an intelligence community assessment was produced within two months of the vote, with an unclassified summary being made public". Stewart Hosie of the Scottish National Party told the media that: "There has been no assessment of Russian interference in the EU referendum, and this goes back to nobody wanting to touch this issue with a 10-foot pole".[45] Britain's Parliamentary Committee on Intelligence and Security concluded that the UK, in contrast to the United States, had made no effort to establish how much Russia may have interfered with the 2016 Brexit referendum.[46] Indeed the US government was unequivocal: "The Committee found that the Internet Research Agency sought to influence the 2016 presidential election by harming Hillary Clinton's chances of success and supporting Donald Trump at the direction of the Kremlin... Russia's goals were to undermine public faith in the US democratic process, denigrate Secretary Clinton and harm her electability and potential presidency".[47]

However, in absence of a firm (or at least, public) evidence base, the case against Russia will continue to be centred on people's opinions – giving Russia and her supporters plenty of room to sow doubt. But much of the intelligence and research community have formed fairly definite opinions. Take former UK National Security Advisor, Sir Peter Ricketts: "Following the 2016 EU referendum there was a considerable and growing body of evidence that Russia was interested in and acting to disturb Western election

---

[45] Mackinnon A, *4 Key Takeaways From the British Report on Russian Interference*, Foreign Policy Journal, 21 July 2020, available at https://foreignpolicy.com/2020/07/21/britain-report-russian-interference-brexit/.
[46] Ibid.
[47] *Report of the Select Committee on Intelligence United States Senate on Russian Active Measures Campaigns and Interference in the 2016 U.S. Election: Volume 2: Russia's Use of Social Media, with Additional Views*, available at https://digitalcommons.unl.edu/senatedocs/4/.

processes … It is a matter of public record that both the U.S. election in 2016 and the French election in 2017 were subject to Russian interference".[48] Across the community it is hard to find anyone who really disagrees; if there is a point of contention it is to presume that Russia interfered only on the side of supporting Brexit. Some eminent Russian experts have coalesced around a view that Russia did indeed interfere in the referendum but, intriguingly, may well have done so on both sides.[49]

This may seem counter intuitive. The chaos and polarisation that the Brexit referendum caused in UK society, even if it had not led to a leave vote, is more than enough to have met Russia's aims at destabilising and undermining British governance. And British politics has become deeply unstable. Since the 2016 referendum to publication there have been five UK Prime Ministers, two leaders of the official opposition and four leaders of the Liberal-Democrats. The Conservative Party had dramatically tilted to the political right with many of its centralists expelled. The UK Parliament was unlawfully prorogued in September 2019 leading to accusations that the executive was circumnavigating democratic governance. The Far Right has seen a resurgence in mainstream UK political life and the EU, the UK's largest trading partner, has now in many respects been demonised by the elected government of the UK, undoubtedly encouraging similar anti-EU factions in Italy, Holland, and France. All of that would meet with President Putin's approval.

It is clear that in the Brexit referendum a complex series of issues came together to create a perfect storm. Perhaps in part this is why the fixation on Cambridge Analytica has been so compelling; liberal Britain (of which I include myself) has been desperate to find a 'bad guy' to blame and Cambridge Analytica and Nix were in the wrong place at the right time. The reality of the Cambridge Analytica affair is that they, and in particular Nix, were astonishingly good at marketing themselves. The affair showed how poorly population groups were understood and the arrogance of liberal elites who were so utterly shocked at both the Trump election and the Brexit result that they searched desperately for someone to blame. The affair showed how even the most passionate investigative journalism needs to be tempered with critical thinking.

---

48 Jukes P, *Landmark Ruling in Strasbourg as MPs Challenge UK Government over Failure to Investigate Russian Interference in Brexit*, Byline Times, 19 January 2023, available at https://bylinetimes.com/2023/01/19/landmark-ruling-in-strasbourg-as-mps-challenge-uk-government-over-failure-to-investigate-russian-interference-in-brexit/.
49 Private unclassified academic security and intelligence discussion forum July 2022.

There are undoubtedly more questions to be asked. I cannot help but feel that some of them ought to be about the very energetic 'witchfinders', of whom Dr Emma Briant, who refers to herself as 'the maven of persuasion'[50] has perhaps been the most vocal. As "an internationally recognised expert and scholar of information warfare and propaganda"[51] and a self-titled 'expert' on Cambridge Analytica, she had had significant access to SCL Defence long before the scandal erupted. It is therefore disappointing to me, who facilitated that access, that she seemed to me unable to differentiate between the hype and actual evidence of Cambridge Analytica's political work, and what she knew SCL Defence was doing for military clients. Even the original journalist who exposed Cambridge Analytica, Carole Cadwalladr of The Guardian newspaper, wrote to me in 2018 and told me that: "I can see that there's an important distinction to be made between SCL Defence's work [and Cambridge Analytica]".[52]

Dr Briant has written widely about the ethics of Target Audience Analysis and defence-based research; yet I wonder if it is at all possible that she may have been remiss in her own conduct. Universities have strict ethical guidelines that govern how different type of research can be undertaken and I wondered if she had complied with her own University's ethical guidelines on research. I wrote to the University of Essex to find out. Despite an apparently exhaustive review process conducted by the Pro-Vice Chancellor for Research, the University was unable to confirm to me one way or the other if Dr Briant had indeed complied with their ethics guidance.[53] Certainly, Dr Briant's decision to make public her privately recorded interviews with Nigel Oakes, whom can be clearly heard asking (and being reassured) that his comments were off record and would not be made public, caused unease amongst some academics. The AHRECS website, which states "that it is a voice for constructive change ... in human research ethics"[54] opined that "Dr Briant has since faced suggestions that her

---

50 http://emma-briant.co.uk.
51 Ibid.
52 Email Cadwalladr / Tatham dated 25 February 2018.
53 Between January and May 2020 I made a series of Freedom of Information requests to the Essex University on this issue – none elicited the definitive answer I requested. The final correspondence I received, on the 21 May 2020, apologised for the delay and promised to reply more fully in due course. The Freedom of Information Team, Innovation & Technology Solutions at the University of Essex told me that *"Please be assured that we haven't forgotten about your enquiry. We have allocated time to look into your enquiry and once we have done this we will come back to you as soon as we can"*. Some three years later, and despite follow on letters to the Chancellor of the University, I am still waiting.
54 https://ahrecs.com/about-us/.

decision to hand over the material raises questions about research ethics, particularly in the context of the Economic and Social Research Council's principles of ethical research".[55]

Dr Briant seems quite closely connected to Professor David Miller, whom we met in the previous chapter and who has done much via his Powerbase and Spinwatch websites and his @Tracking_Power Twitter (now 'X') account to draw attention to IO. Dr Briant, who studied alongside Professor Miller, told me in 2016 that he was 'a close personal friend'. Indeed Dr Briant has joined Professor Miller in signing letters to various newspapers including one in which they asserted that the UK's counter-terrorism policy, PREVENT, "must be abolished and firmly uprooted from society",[56] and another in which they jointly signed a statement stating that "Western governments have ... fuelled and facilitated groups such as ISIS".[57] Miller was dismissed from his job at Bristol University for alleged anti-Semitism, a claim he contested and was subsequently cleared of. His Twitter (now 'X') account was apparently once labelled as 'Iran State Affiliated Media'[58,59] and he has certainly been a regular guest on Iran's Press TV. Early in 2023, he told Press TV that "Soleimani is somebody who, during his life, accomplished military feats, for which the whole world should be grateful".[60] This is the same General Soleimani who oversaw, before his death, the Iranian Republican Guard who in 2020 shot down a civilian airliner,[61] accused of 'enforced disappearances'[62] of Kurdish minorities'

---

55 *Unethical not to' submit Brexit interviews to MPs, says academic*, AHRECs, 20 May 2018, available at https://ahrecs.com/latestnews/uk-unethical-not-to-submit-brexit-interviews-to-mps-says-academic-the-john-morgan-april-2018/.
56 *Towards a Society Beyond PREVENT* available at http://www.protectingthought.co.uk.
57 *24th November 2015 – Letter to The Guardian following recent attacks*, available at https://emma-briant.co.uk/24th-november-2015-letter-to-the-guardian-following-recent-attacks/.
58 See https://twitter.com/daverich1/status/1540626193225064448?lang=en.
59 Twitter defines accounts that are labelled as being 'state affiliated' as: outlets where the state exercises control over editorial content through financial resources, direct or indirect political pressures, and/or control over production and distribution. See https://help.twitter.com/en/rules-and-policies/state-affiliated.
60 *Gen. Soleimani 'unifying figure', 'architect of resistance'*: Academic, Press TV, 1 January 2023, available at https://www.presstv.ir/Detail/2023/01/01/695522/Gen-Soleimani-unifying-figure-architect-of-resistance-Academic.
61 *Iran Says It Unintentionally Shot Down Ukrainian Airliner*, New York Times, 10 January 2020, available at https://www.nytimes.com/2020/01/10/world/middleeast/missile-iran-plane-crash.html.
62 Loft P, *Iran protests 2022: Human rights and international response*, UK Parliament, available at https://researchbriefings.files.parliament.uk/documents/CBP-9679/CBP-9679.pdf.

and 'torture'[63] and which has allegedly attempted ten assassinations in the UK.[64] Miller has also spoken of: "strong confluence…between Zionist ideas and Ukrainian nationalist ideas. These are far-right ideas … Zelensky himself he's Jewish, he's also strongly oriented towards Israel, but that in itself is a sign of being oriented really towards the far-right".[65] Miller has also tweeted that "Jews are not discriminated against" and that "they are over represented in positions of cultural, economic and political power".[66] All of which prompted the Alliance for Securing Democracy at the German Marshall Fund to conclude that: "In Miller … Tehran has found validators who will not only launder their disinformation but build and expand on it in new and destructive ways".[67]

Given Briant and Miller's shared areas of interest in propaganda, IO and government 'spin', and their friendship, it would not be unusual if Briant's and Miller's research interests have not coalesced in recent years. She has, for example, submitted Freedom of Information requests to find out what has been said about Miller at the British Army's 77th Brigade.[68] Indeed, she is an enthusiastic submitter of such requests. A search of Whatdotheyknow.com suggests that could be around 300 via that website and there may be more made directly[69], which seems at odds with her proclivity for ethical and consent-based research, given that it compels public servants to 'participate' in her research whether they wish to or not. Yet for an expert in propaganda, her website seems remarkably devoid of any examination of current Russian, Chinese or Iranian propaganda activities and active measures, which collectively amount to probably the biggest concerted propaganda campaigns ever mounted. Neither is there any obvious research into historical propaganda. Indeed her website lists

---

63 *Tortured Kurdish journalist taken to hospital in Iran's Sanandaj*, Kurdistan Human Rights Network, 16 January 2022, available at https://kurdistanhumanrights.org/en/tortured-kurdish-journalist-taken-to-hospital-in-irans-sanandaj/.
64 Mendick R, *Iran tried to assassinate British residents 10 times this year, MI5 chief reveals*, Daily Telegraph, available at https://www.telegraph.co.uk/news/2022/11/16/iran-tried-assassinate-british-residents-10-times-year-mi5-reveals/.
65 Kohlenburg N, *British Commentators Promote Pro-Russian Conspiracy Theories on Iranian State-Backed Media*, Securing Democracy, 20 October 2022, available at https://securingdemocracy.gmfus.org/british-commentators-iran-russia-ukraine-conspiracy-theories/.
66 @tracking_power Twitter (Now 'X") 6 August 2023 11.07pm.
67 Ibid.
68 https://www.whatdotheyknow.com/request/670267/response/1602501/attach/html/3/20200716%20FOI06720%20Briant%20PIT%20Extension%20Letter%20ArmySec.pdf.html.
69 Search of Whatdotheyknow.com results availabe at https://www.whatdotheyknow.com/search/emma%20briant/all/newest?query=emma+briant

her contribution to nine edited books (not one appears to cover Russian or Chinese IW), to four books of her own (again, none appear to cover Russian or Chinese propaganda) and six academic papers of which only one appears to touch on Russian propaganda activity.

For the advancement of military IO, the Cambridge Analytica affair was a disaster. SCL Defence had undertaken good work for both the US and UK militaries; people across Defence were beginning to understand that social science was the basis for effective and persuasive communication to mitigate bad and encourage good behaviours, and not for creative marketing inspired adverts. But overnight the SCL brand became completely toxic and past and current clients ran as fast as they could from the story. Senior officers and IO practitioners who had worked in the MoD on the BD methodology, who (alongside me) had seen its merit and shouted its praises, were suddenly, and probably understandably, quiet. From Canada a long-time friend and military strategic communication's specialist, texted me to say that: "in my work here with Chief of Defence, I can see real trepidation of senior military and political leaders now when they see any file associated with IO and PsyOps – this is an outcome that is fallout from association with SCL".[70]

The NATO Centre of Excellence for Strategic Communications in Latvia has been the recipient of SCL Defence's TAA training. Its director, Jānis Sarts, told Latvian media there was nothing "mega-new" in the content of the course, which took place three years ago and that the centre had not been particularly impressed by the content of the course or the value for money offered. Subsequently the services of SCL or Cambridge Analytica have not been used.[71] This appears at odds with the sign off report submitted by Mr Sarts to the Canadian government on completion of the CAN$1million BD training course undertake at the COE over 8 weeks. Sarts wrote that: "with 20 trained BD TAA specialists resulting from this course and of these 19 certified to conduct further training in their own nations, the StratCom COE and its staff, its sponsoring nations and allied (Canada and USA) and partner (Finland) nations has significantly increased their capacity in TAA. This is further amplified by the professional relationship formed over the course – with SCL staff, visiting academics, subject matter experts and fellow classmates. The network established amongst those

---

70 Message via LinkedIn 3 Feb 2021.
71 *Cambridge Analytica linked company taught at NATO STRATCOMCOE, 26 March 2018, available at* https://eng.lsm.lv/article/society/society/cambridge-analytica-linked-company-taught-at-nato-stratcomcoe.a272688/.

who participated will be important to joint efforts now and, in the future, to inform NATO".[72] Far from not being used, SCL Defence accompanied the NATO COE delegations to Georgia, Ukraine and Moldovia to assist those nations to understand how pervasive influence was being used by Russia to undermine democratically elected governments and institutions.

The Cambridge Analytica story destroyed that progress almost overnight and it became the perfect exemplar of what former UK Foreign Secretary, Douglas Hurd, once described as "the light shown by the media is not the regular sweep of a lighthouse, but a random searchlight directed at the whim of its controllers".[73] The searchlight picked out Cambridge Analytica and Alexander Nix and with the enthusiastic help of Dr Briant it picked out SCL Defence, military grade IO and Target Audience Analysis. Its sweep blinded the MoD, the DOD and NATO as senior officials panicked and ran from its glare.

But the searchlight did not pick out M&C Saatchi, and its Director, former 15 (UK) PsyOps Army Officer, Simon Bergman,[74] and their huge contracts with the US DOD and RICU; it did not pick out the Adam Smith Institute who won a £30 million contract to deliver behavioural change for RICU; it did not pick out NSI in the US, led by Dr Robert Popp former Assistant Deputy Undersecretary of Defense for Advanced Systems and Concepts, which according to its website helps "clients understand people and their behaviors thru multidisciplinary data-driven analytics", it didn't pick out former UK Army Officer, Paul Tilley, and his company IN2, which provides "communication interventions .. informed by in-depth research or local insight (on the ground and online) to ensure our messages and activities resonate with our audiences and add value to our clients".[75] It didn't sweep over the US' Rendon corporation which "through a myriad of customized and comprehensive research and open-source intelligence products, TRG provides its global clients with focused situational awareness and insight into the geopolitical landscape needed to improve operations".[76] The light

---

72 Final project Report submitted by Janis Sarts to Canadian Stabilisation and Reconstruction Task Force dated 4 November 2015. Available as a PDF on www.IOFFC.info.
73 Hudson M and Stanier J, *War and the Media,* Sutton Publishing, 1999.
74 Associated Press, *How the US military is botching the online fight against ISIS due to incompetence, cronyism flawed data, and a lack of Arabic speakers,* The Daily Mail, 31 January 2017, available at https://www.dailymail.co.uk/news/article-4174644/US-military-botches-online-fight-against-Islamic-State.html.
75 Personal LinkedIn page of Paul Tilley MBE available at https://www.linkedin.com/in/paul-tilley-mbe-2972056/?originalSubdomain=uk.
76 The Rendon Corporation website available at https://rendon.com.

missed Artifice, which on its website states that: "Changing the behaviour of your enemies and competitors in an adversarial context is not an easy challenge. However, it's not as though you have a choice… and we can give you the tools you need to meet that challenge",[77] or Leonie Industries which specialises "in reaching target audiences in challenging locations".[78] It failed to pick out Sayara International that: "uses rigorous research and communication to create positive behaviour change in fragile and conflict areas".[79] All of this was invisible from the picture painted by the global media coverage and the expertise of Dr Briant.

There are two terms which are often used to describe how media cover issues – 'priming' and 'framing'. The first, Priming, is used to define what the media wants us to think about. The second, framing, tells us how we should think about it. It was perfectly reasonable of the media to report on Cambridge Analytica (priming); but the manner of that reporting – the framing – provided readers with little or no context to understand the story. In sensationalising Cambridge Analytica, which remember had no military access at all, The Guardian and Dr Briant (the self-styled 'maven of persuasion') told the world's population, in my opinion, that this was something unique, unusual and was the use of military techniques to subvert democratic processes; that was not the case.

Alongside the NATO Centre of Excellence, the Dutch and Canadian Armed Forces had both bought BD training from SCL Defence as the basis for their IO. Whilst I played no role in the Canadian programme I had spent almost three years working with Dutch colleagues and was pleased to hear excellent reports from their Special Forces, in particular, about the use of BD in their military mission to Mali to protect the population against improvised explosives. Indeed the Dutch Minister of Defence, Dr A.Th.B. Bijleveld-Schouten,[80] reported to Parliament that after research with the Behavioral Dynamics Methodology: "target group-oriented communication was adapted to local circumstances. After this, locals reported more IED

---

77 Artiface website available at https://artifice.co.uk.
78 Leonie Industries detailed at https://www.devex.com/organizations/leonie-industries-46377.
79 Sayara International website available at https://sayarainternational.com.
80 Beantwoording Kamervragen over het artikel 'Een soft maar gevaarlijk wapen: moderne oorlogsvoering richt zich op beïnvloeding van de bevolking'. 24 August 2020, reference: BS2020014218, published on the Dutch Government website: https://www.rijksoverheid.nl/documenten/kamerstukken/2020/08/24/beantwoording-kamervragen-over-gedragsbeinvloeding-in-conflictgebieden.

sites that could then be rendered harmless. This method therefore broadens the range of instruments of Defense and makes Defense more effective".

In June 2020, as a response to media coverage, some Dutch members of parliament submitted questions to the minister of Defence which were duly answered. A full list of the questions is provided at the reference[81] but some are worth repeating in full: Question. Is it true that the BDM method was developed by the parent company of Cambridge Analytica and purchased by the Ministry of Defense? Why was the House never informed about the collaboration with this controversial company? Answer. The Ministry of Defense did not purchase the BDM course from Cambridge Analytica in 2017, but from SCL Defense…the use of big data was not part of the training purchased by Defense. Furthermore, Defense considers BDM as an instrument and the ethics of the instrument resides in the user and not in the instrument. Unlike the UK, and Canada, and the NATO Centre of Excellence, the Dutch have continued to use the BD process regarding it as invaluable to their operations, telling their parliamentarians that "the Netherlands shares the view that the protection of and cooperation with civilians in conflicts should be central (people and behavior centric). BDM fits well with this approach. The Netherlands are at the forefront in the operationalization of BDM in military operations and deployment. It is significant that other countries are knocking on the door of our armed forces during the pilots to learn from us".[82]

In Eastern Europe, in North Africa, across the recruiting grounds of ISIS and AQ, the BD methodology worked successfully every time it was used in IO. Whilst Cambridge Analytica was nothing new, innovative or amazing, SCL Defence's BD methodology was. But in the fall out from the Cambridge Analytica story the defence capability was completely and undeservedly discredited and ultimately, in the UK and NATO, destroyed. That it no longer exists, and that the BD variant of Target Audience Analysis has been so demonised globally, is a source of bitter personal regret. Our collective IO armoury is emptier and less effective as a result and our understanding of adversaries IO, and their effects on vulnerable populations, now significantly reduced.

---

81 See www.IOFFC.info.
82 Beantwoording Kamervragen over het artikel 'Een soft maar gevaarlijk wapen: moderne oorlogsvoering richt zich op beïnvloeding van de bevolking'. 24 August 2020, reference: BS2020014218, published on the Dutch Government website: https://www.rijksoverheid.nl/documenten/kamerstukken/2020/08/24/beantwoording-kamervragen-over-gedragsbeinvloeding-in-conflictgebieden.

12 The Future?

The writer, author and journalist Dorian Linskey wrote in the UK's Guardian newspaper in 2019 that: "the growth of "deep fake" image synthesis, which combines computer graphics and artificial intelligence to manufacture images whose artificiality can only be identified by expert analysis, has the potential to create a paranoid labyrinth in which, according to the viewer's bias, fake images will pass as real, while real ones are dismissed as fake".[1] It may be a dystopian view but is it realistic?

In October 2022, my wife and I went to see ABBA in concert. The 'real' ABBA had last given a concert on 11th December 1982; what we went to see were holographic recreations of the four singers in a purpose-built arena in East London. Looking exactly as they did in the mid-1970s, the holographs were remarkably good – from a distance they were indistinguishable from real people but on the big screens and in close up the faces were just very slightly wooden – and we all knew that they were just very clever IT, the result of £40 million to develop and three years of work. But what we had not expected was what happened at the concert's end. As the 1970's figures faded from view, the ABBA of today (four 70 plus year old singers) walked on from the side of the stage, waving, and thanked us all for coming. To this day I have no idea if they were the real, human, ABBA (which seems highly unlikely given how many performances each week were taking place) or just more stunningly realistic holographs.

Why is this anecdote relevant to IO? The answer is that it is the latest, and probably most sophisticated steppingstone so far on the Fake News journey. Whilst it may be true that mainstream news media in western nations has become more proactive in filtering out doctored photographs, it is by no means fool proof; online media outlets however have been far less rigorous and fake imagery perpetuates, be it accidental misinformation or deliberate disinformation. A paper written for the National Library of

---

[1] Lynskey D, *Nothing but the truth: the legacy of George Orwell's Nineteen Eighty-Four*, The Observer, 19 May 2019, available at https://www.theguardian.com/books/2019/may/19/legacy-george-orwell-nineteen-eighty-four.

Medicine by psychologists at the UK's Warwick University concluded that: "People exhibit a bias to accept images as "real," so one-third of manipulated photos go undetected" and that "the ability to distinguish real from fake photos declines with age".[2] In 2017 the BBC reported that fake video had now reached such quality that it was almost impossible to detect. They reported that researchers at the University of Washington had managed to create a video of former US President Barack Obama making a faked speech. In particular they noted how Artificial Intelligence (AI) has been used to precisely model how President Obama moved his mouth when he spoke.[3] Their techniques potentially allowed them to put any words they wanted into their synthetic president's mouth. In 2018 Bloomberg ran a report on fake video, showcasing the work of Jordan Peel who, again, had chosen Obama for his fakery. The video also showed how the same technology could be used to convincingly fake video of changing weather patterns – something that would be very useful to both climate deniers and climate activists. One contributor described the technology as "deep fakes on steroids" and suggested that it could very easily be used to damage reputations or for individuals who had said objectionable things in the past to claim that their words had been faked by the technology.[4] In October 2023 US President Joe Biden announced that he would issue an executive order that would impose safety controls on future AI. It came after the President saw 'fake AI images of himself, of his dog ... and he's seen and heard the incredible and terrifying technology of voice cloning, which can take three seconds of your voice and turn it into an entire fake conversation."[5]

Facilitating the distribution of fake material is social media. As a consequence, the number and range of tools that have been developed to monitor, assess and directly intervene in that environment is significant and it appears as if almost all effort is now being channelled into the many different social media channels – notably the 'big five' of Facebook (and its lookalikes such as Vk in Russia), Twitter, Instagram, TikTok and YouTube.

---

2 Nightingale SJ, Wade KA, & Watson DG, *Can people identify original and manipulated photos of real-world scenes?* Cognitive Research: Principles and Implications, 2. doi: 10.1186/s41235-017-0067-2.
3 *Fake Obama created using AI tool to make phoney speeches*, BBC News, available at https://www.bbc.co.uk/news/av/technology-40598465.
4 *It's Getting Harder to Spot a Deep Fake Video*, available at https://www.youtube.com/watch?v=gLoI9hAX9dw.
5 Kilander G, *Why Biden is so concerned about AI*, The Independent 2 November 2023, available at https://www.independent.co.uk/news/world/americas/us-politics/joe-biden-ai-executive-order-b2440366.html.

That effort is both offensive – proactively messaging audiences via the social media channel through overt and covert means – and to a lesser extent defensive in trying to track and guard against foreign offensive efforts such as those coming out of the famous St Petersburg Troll Farm described in the chapter on Russia.

For all the sophistication of the technology, is it right to place such a huge focus on social media? With so many fake accounts and personas will 'our' bots spend their time messaging 'their' bots, and vice versa. Or put another way, how on earth do we know who we are speaking to? And how do we actually measure the success of our online campaigns if we have no idea with whom we are communicating? Will we revert to easy but ultimately misleading metrics – so called measures of performance such as number of shares, number of likes and number of comments? It seems likely.

And what about conflict in nations where we cannot permeate organic social media – such as North Korea – or where is just doesn't exist? Whilst the global literacy rate is high, some 86 per cent, there are many nations where that is not the case – Chad, Mali, Ethiopia, Afghanistan, Sudan, and Mauritania for example. These may not be global superpowers, but they are countries where western forces have had to intervene or may yet do so again. Overlay internet access and social media usage with literacy rates and it becomes apparent that the entire world is not holding a smart phone or sat in front of a personal computer. As of January 2023, 93 per cent of the population in South Sudan did not have internet access. Somalia and Burundi followed with 90 per cent of the population offline. North Korea has zero per cent penetration as the World Wide Web remains blocked for its citizens. And it would not be difficult for other nations to turn off internet access for their citizens. Myanmar, for example, has done so a few times[6] and a recent report suggests that in 2021 Governments intentionally shut down internets 182 times across 34 countries.[7] In Mali, a very troubled nation, French and Dutch troops operating with the UN struggled to find relevant conduits to reach a population with 30 per cent literacy and less than 30 per cent internet penetration (and then most of that not in the

---

6 Ratcliffe R, *Myanmar coup: military expands internet shutdown*, The Guardian, 2 April 2022, available at https://www.theguardian.com/world/2021/apr/02/myanmar-coup-military-expands-internet-shutdown.
7 Greig J, *Governments intentionally shut down internet 182 times across 34 countries in 2021: report*, The Record, 28 April 2022, available *at* https://therecord.media/governments-intentionally-shut-down-internet-182-times-across-34-countries-in-2021-report.

areas of conflict but in the major cities).[8] If the future, primary, means of IO message distribution is based mainly on using social media channels it seems likely there will be a host of potential future audiences who will be missed.

Offensive IO on Social Media

It was Goebbels, Hitler's infamous propagandist, who so famously commented on the desirability of repetition to spread messages: "Repeat a lie often enough and it becomes the truth".[9] In psychology this is often referred to as the illusory truth effect and is thought to occur because repetition increases the brain's processing fluency.[10] We also see this has a direct read across to the BD methodology's principle of normative affiliation. If the vast mass of people around you (and that includes those commenting on the social media posts that you read) are adopting a specific view or position it is very compelling to take the same view, even if you initially disagree. This is why volume is so important and in particular why so many automated accounts – bots – exist in the information environment because you need a great many posts in order to achieve normalisation. One of the ways this might be done is via persona management and disguised attribution.

In September 2022, The Washington Post reported that the US Department of Defense had ordered an investigation after various social media companies had identified, and subsequently taken offline, fake accounts suspected of being run by the US military.[11] Internet researchers Graphika and the Stanford Internet Observatory had identified over 150 bogus personas operating promoting pro-Western narratives in the Middle East and Central Asia information environment. Graphika reported that: "These campaigns consistently advanced narratives promoting the interests of the United States and its allies while opposing countries including Russia, China, and Iran. The accounts heavily criticized Russia in particular for the deaths of innocent civilians and other atrocities its soldiers committed in pursuit of the Kremlin's "imperial ambitions" following its

---
8 Author's discussion with recently retired Dutch Armed Forces IO officer.
9 Stafford T, *How liars create the 'illusion of truth'*, The BBC, 26 October 2016, available at https://www.bbc.com/future/article/20161026-how-liars-create-the-illusion-of-truth.
10 Hasan A & Barber S, *The effects of repetition frequency on the illusory truth effect*, Cogn. Research 6, 38 (2021) available at https://cognitiveresearchjournal.springeropen.com/articles/10.1186/s41235-021-00301-5.
11 https://www.washingtonpost.com/national-security/2022/09/19/pentagon-psychological-operations-facebook-twitter/.

invasion of Ukraine in February this year".[12] The Twitter dataset alone numbered nearly 300,000 tweets spanning the ten years between March 2012 and February 2022.

In March 2011, the media reported that US IT company NTrepid had won a $2.76 million contract from the US military for "online persona management".[13] In 2015 the media reported that the U.S. Air Force had asked the same US company to create software that would enable it to mass-produce bots for messaging purposes.[14,15]

On 15th February 2023, a group of newspapers simultaneously published an expose of a team of Israeli contractors who claimed to have manipulated 30 presidential elections globally. Allegedly run by former Special Forces soldier, Tal Hanan, who went by the pseudonym Jorge, the newspapers reported that one of his services was a software package called Advanced Impact Media Solutions (AIMS), that was allegedly able to control thousands of fake social media profiles on Twitter, LinkedIn, Facebook, Telegram, Gmail, Instagram, and YouTube.[16]

In March 2022, the UK MoD posted a public tender document to the web for Online Engagement Services (OES). The document stated that the MoD had a requirement for a: "System Innovator/Integrator (SII) to deliver a Cloud based online influence and engagement service". The tender was split against two core services: "A persona management capability to create & maintain live personas for use with online services such as social media" and "IT infrastructure & application services, utilising managed attribution. This includes but is not limited to social media, online surveys and other methods",[17] all of which sounds very similar to the 2011 and 2015 contracts posted by the US DOD.

In February 2023, Meta announced that its Facebook and Instagram platforms would require users to pay to be verified on the social media platforms. Users would need to show approved ID, their accounts

---

12 https://public-assets.graphika.com/reports/graphika_stanford_internet_observatory_report_unheard_voice.pdf.
13 Nick Fielding and Ian Cobain, "Revealed: US spy operation that manipulates social media", *The Guardian*, March 17, 2011. Available at https://www.theguardian.com/technology/2011/mar/17/us-spy-operation-social-networks
14 https://www.forbes.com/sites/lutzfinger/2015/02/17/do-evil-the-business-of-social-media-bots/?sh=7971756efb58.
15 https://www.telegraph.co.uk/technology/social-media/8389577/Pentagon-buys-social-networking-spy-software.html.
16 https://www.theguardian.com/world/2023/feb/15/revealed-disinformation-team-jorge-claim-meddling-elections-tal-hanan.
17 https://bidstats.uk/tenders/2022/W12/771358873.

would have to have a posting history and users must be at least 18 years old.[18] The move followed the announcement from Twitter that users could henceforth pay for their accounts to be verified using a paid for SMS-based two-factor authentication system. Both are an attempt by the social media companies to clamp down on the huge numbers of fake accounts that exist on their respective platforms, and, of course, to monetise their usage further. In the third quarter of 2022, Facebook took down over 1.5 billion fake accounts, an increase of 100 million on the previous quarter. A record figure of approximately 2.2 billion fake profiles were removed by the platform in the first quarter of 2019.[19] Amongst those were the US government accounts. As Facebook told The Washington Post, "The accounts were easily detected by Facebook, which since Russia's campaign to interfere in the 2016 presidential election has enhanced its ability to identify mock personas and sites. In some cases, the company had removed profiles, which appeared to be associated with the military, that promoted information deemed by fact-checkers to be false".[20]

However, in the cat and mouse environment of social media and online presence, new tools are quickly developed; take MODLISHKA for example,[21] a piece of code that can be downloaded to by-pass the need for two-factor authentication of accounts. As the social media companies attempt to impose more security, so hackers and coders will do their best to subvert. The desire to influence in social media, particularly covertly and at volume, is likely to become not just a major technological challenge, but the next battleground in IO.

Social Media Analysis

The number of companies providing social media analysis is astonishing and presents some serious challenges. Accessing social media channels is not in itself difficult, the problem of course is how to sort the huge tsunami of data to find meaningful content. And so increasingly companies will offer ever more sophisticated tools to monitor social media channels. The

---

18 https://www.theguardian.com/technology/2023/feb/20/facebook-instagram-paid-meta-verification-twitter-charge-2fa-two-factor-authentication-via-sms.
19 https://www.statista.com/statistics/1013474/facebook-fake-account-removal-quarter/.
20 Nakishema E, *Pentagon opens sweeping review of clandestine psychological operations,* The Washington Post, 19 September 2022, available at https://www.washingtonpost.com/national-security/2022/09/19/pentagon-psychological-operations-facebook-twitter/.
21 Fireshark, *Phishing Attacks! How Modlishka Works & How to Set up Modlishka in Kali Linux | Ethical Hacking,* available at https://www.youtube.com/watch?v=6CSPVOIfVCY.

Pulse tool,[22] developed by Two Six Technology in the US is a good example and is now widely used by a number of US military Combatant commands. The systems allow users to home in on specific areas of discussion and if necessary to engage directly with some of the key influencers in the debate. The two problems that all such tools have is the sheer volume of data (and it is worth remembering that in Afghanistan, one ISAF intelligence organisation was collecting a terabyte of data a day) and secondly, how certain are we that the data is 'real'? Does it originate from humans? Does it originate from the humans it purports or is it just machine manufactured? (a modern-day equivalent of the conundrum faced by McNeil industries in the Helmand Perception Matrix discussed in the chapter on Afghanistan earlier). Social media tools are likely to get a whole lot smarter, but they will need to because so will everything else and at the moment some of the smartest players are non-state actors.

Non-State Actors

It is not just the Russians who are being caught out by social media. The urge to film and then share events was the downfall of so called Marine A, Sgt Blackman, who in Afghanistan in 2011 shot dead an insurgent who had been seriously injured in an attack by an Apache helicopter. Footage of the incident, recorded on a colleague's helmet mounted camera and, as is human nature, shared amongst those present, was discovered the following year during an unrelated police investigation, and ultimately resulted in Blackman's conviction and a lengthy prison sentence But the intentional 'outing' of violent and often illegal behaviours has become an industry all of its own and the undisputed leaders are Bellingcat and Anonymous.

Anonymous is a decentralized international activist and hacktivist collective and movement primarily known for its various cyberattacks against several governments, government institutions and government agencies. Weeks after declaring an "electronic war" on the "Kremlin's criminal regime", Anonymous posted on Twitter[23] that they had 2,500 Russian and Belarusian government, state media and other sites "in support of Ukraine" releasing names and locations of units onto the web. Some weeks later Anonymous struck again by announcing on

---

22 https://twosixtech.com/products/pulse/.
23 @YourAnonTV, *Anonymous announcement*, Twitter, 17 March 2022, available at https://twitter.com/YourAnonTV/status/1504556362960879616.

Twitter that it had successfully breached and leaked the personal data of 120,000 Russian soldiers,[24] including names, dates of birth, addresses, unit affiliation and passport numbers.[25] Anonymous also claims to be behind the hacking of several Russian television channels, ingeniously replacing their programming with coverage of the war in Ukraine by two independent broadcasters banned in Russia; Current Time and Dozhd TV.[26] Other hacks attributed to Anonymous include a yacht allegedly belonging to Putin had its electronic call sign changed to "FCKPTN" and its heading setting changed to "hell"[27] and the hacking into unsecured WIFI enabled printers in offices around Russia to spread anti-propaganda messages.[28] The included a warning that Putin, the Kremlin and the Russian media were lying about the invasion and gave instructions on how to access a browser that would allow Russian citizens to bypass the country's censorship. The collective also claimed responsibility for hacking the Russian owned Yandex Taxi company in early September 2022, sending dozens of cars to a location resulting in a traffic jam that lasted up to three hours.[29] Athina Karatzogianni, a media and communications lecturer at the UK's University of Leicester, told FRANCE 24 TV that "There has never been such a mobilisation of hacktivists at the international level to defend the same cause".[30]

Founded by British blogger, Eliot Higgins, Bellingcat has become one of the most globally known and successful open-source analyst companies. Be it the unmasking of Syrian war criminals or Russian hit squad, Bellingcat

---

24 Chirinos C, *Anonymous takes revenge on Putin's brutal Ukraine invasion by leaking personal data of 120,000 Russian soldiers*, Fortune, 4 April 2022, available at https://fortune.com/2022/04/04/anonymous-leaks-russian-soldier-data-ukraine-invasion/.
25 Stanton A, *Anonymous Apparently Behind Doxing of 120K Russian Soldiers in Ukraine War*, Newsweek, 3 April 2022, available at https://www.newsweek.com/anonymous-leaks-personal-data-120k-russian-soldiers-fighting-ukraine-1694555.
26 Russian TV Channels Hacked To Show Independent Coverage Of War In Ukraine, Radio Free Europe, 7 March 2022, available at https://www.rferl.org/a/russian-tv-hacked-ukraine-anonymous/31740663.html.
27 Dellinger A, *Hackers Set The Intended Destination Of Putin's $100 Million Yacht To "Hell"*, MIC, 2 March 2022, available at https://www.mic.com/impact/anonymous-hackers-putin-yacht-call-sign.
28 Kika T, *Anonymous Hacks Into Russian Printers to Deliver Resistance Information*, Newsweek, 21 March 2022, available at https://newsweek.com/anonymous-hacks-russian-printers-deliver-resistance-information-1690269.
29 Papadopoulos L, *A hacker attacked Yandex Taxi and sent dozens of cars to the same location*, Interesting Engineering, 2 September 2022, available at https://interestingengineering.com/culture/hacker-attacked-yandex-taxi-moscow.
30 Seibt S, *Ukraine conflict presents a minefield for Anonymous and hacktivists*, France24, 23 March 2022, available at https://www.france24.com/en/europe/20220323-ukraine-conflict-presents-a-minefield-for-anonymous-and-hacktivists.

has become expert at using open-source data to track individuals on battlefields globally. Amongst some of its biggest successes has been the identification of the Russian unit that shot down Malaysian airline MH17 over Ukraine; the identification of one of the Skripal family poisoners and the work in connecting the poisoning of Russian opposition leader, Alexei Navalny, to the Russian secret service. Its staff have rather proudly stated that the organisation is "the Kremlin's biggest nightmare".[31] There may be an element of hyperbole in that but clearly Bellingcat has been able to expose elements of Russian operations in a way that has not been possible for western nations previously. Whilst this is hugely helpful in supporting the Ukraine position, the West should none the less be wary that Bellingcat's considerable expertise and resource could one day be turned on it, or that Bellingcat spin offs might not have such a positive view of future US or UK military operations.

Artificial Intelligence (AI)

In the 1990s, a series of movies were released starring the US actor Arnold Schwarzenegger as The Terminator. They told the story of human looking machines coming to life and killing people. It made good cinema at the time but in the intervening years many now question if Hollywood fiction has become real life. AI has grown at an exponential pace and the possibility of Terminator like machines, AI powered robots, has not just been questioned but even given a name – The Terminator scenario. How likely is this? Debatable, although Walt Disney recently showcased a prototype robot they had specifically designed to create an emotional connection with people.[32] AI – which rather less threateningly is defined by IBM as "a field which combines computer science and robust datasets to enable problem-solving"[33] – will clearly have significant application in the world of IO. Certainly that is the view of President Putin who in 2017 said that the nation that leads in development of AI would "become ruler of the world".[34]

---

31 *How Bellingcat became Russia's 'biggest nightmare*, France24, 7 September 2022, available at https://www.france24.com/en/live-news/20220907-how-bellingcat-became-russia-s-biggest-nightmare.
32 https://www.linkedin.com/feed/update/urn:li:activity:7040941728756834304.
33 *What is artificial intelligence (AI)?* available at https://www.ibm.com/topics/artificial-intelligence.
34 Strategic Future's Group, *Deeper Looks: The Future of the Battlefield*, March 2021, available at https://www.dni.gov/index.php/gt2040-home/gt2040-deeper-looks/future-of-the-battlefield.

In their 2020 paper *'The Next Generation of Cyber Enabled Information Warfare'*, Keir Giles and Kim Hartman write that the utility of AI will not reside so much in its intrinsic strength, but in the weaknesses of the societies in which it will operate. They note that: "The true power of AI ... derives from several factors: societies' reliance on social media; dependence on cyberspace as a trustworthy information resource; unlimited access to and ability to spread information rapidly through cyberspace; and human difficulties in reliably distinguishing between fake and genuine media".[35] To an extent we are seeing this already. In May 2023 an AI generated image of an explosion at the US Department of Defence quickly went viral and caused a brief dip in the US stock market as investors panicked that the US might be under attack, again. The image showed smoke billowing up from the Pentagon, with media later reporting that "it could be the first instance of an AI-generated image sowing enough confusion to move stock markets".[36]

If we are to navigate future IO, we have to build resilience inside societies to help defend against adversarial influence and an interesting application of AI doing just that is the 'Fifth Column' application being built by Australian Cyber Defence company, Internet 2.0.

Concerned at the number of bots circulating in social media, Internet 2.0 designed Fifth Column,[37] using AI, to learn what malicious bots look like and how they behave. That learning process is ongoing with so far millions of tweets having been scanned by the AI code and that process will continue for some months before the Beta (test) version is released. In their literature, Internet 2.0 state that: "Fifth Column is designed to categorize bot accounts, track group botnet behavior, and collect/record all activity to give the user a world view that they can drill down into".[38] Critically the software will be available in time for the 2024 US Presidential Elections and made available to both Democratic and Republican Parties. There is clearly an altruistic and commendable intent to this programme, but underlying it is the presumption that both sides are likely to invest heavily in automated IO.

Without doubt AI will allow the generating of personalized content to become cheaper and thus more widespread. This personalisation is very

---

35 Giles K & Hartmann K, *The Next Generation of Cyber-Enabled Information Warfare*, CCDOE, available at https://ccdcoe.org/uploads/2020/05/CyCon_2020_13_Hartmann_Giles.pdf.
36 Hurst L, *How a fake image of a Pentagon explosion shared on Twitter caused a real dip on Wall Street*, 23 May 2023, Euronews, available at https://www.euronews.com/next/2023/05/23/fake-news-about-an-explosion-at-the-pentagon-spreads-on-verified-accounts-on-twitter.
37 https://internet2-0.com/5th-column/.
38 Internet 2.0, *'Military Grade Cyber Protection – Fifth Column'* Information Data Sheet provided to author by Internet 2.0.

important. Social media platforms are consistently evolving and developing ways to identify fictitious accounts. Rae Baker, senior OSINT Analyst on the Dynamic Adversary Intelligence team at Deloitte, writes that in the future "we want to create accounts that look as real as possible and often that requires a back story, details about their life, and images of the individual in various poses".[39] To an extent we are already at that point.[40] 2023 saw the first theatre play based solely on a script produced by AI[41]; it was also the year that 'Heart on my sleeve', an AI generated rap song, was removed from streaming services after protests from record labels.

AI will also facilitate the use of different languages. In February 2023 the MixerBox Software company launched MixerBox ChatAI, the world's first artificial intelligence (AI) chatbot Web browser optimized for traditional Chinese.[42] Language is an enduring problem in IO. You need to have very good language skills if you wish messaging to appear organic; AI will likely enable new tactics such as real-time foreign language content generation to emerge.[43] As this book goes to print in 2024, US start-up OpenAI had demonstrated AI powered GPT-4 product capable of not just image recognition but also of making nuanced commentary about that image.[44]

A relatively new term has entered the IO lexicon as a result of advances in AI – Computational Propaganda. This is defined as the "use of algorithms, automation, and human curation to purposefully distribute misleading information over social media networks".[45] Or in other words, the future will be the delivery of more messages, from more believable but none the less fake personas, to ever bigger audiences. That is not an attractive proposition.

---

39 Baker R, *Using AI to Develop Realistic Sock Puppet Accounts*, Rae Baker Blog, available at https://www.raebaker.net/blog/using-ai-to-develop-realistic-sock-puppet-accounts.
40 Mickle, Metz, Grant, *The Chatbots Are Here, and the Internet Industry Is in a Tizzy*, New York Times, 8 March 2023, available at https://www.nytimes.com/2023/03/08/technology/chatbots-disrupt-internet-industry.html.
41 O'Donovan B, *Chatbot play will see Dublin audience take control*, RTE News, 9 March 2023, available at https://www.rte.ie/news/business/2023/0308/1361082-rehearsals-underway-for-artificial-intelligence-play/.
42 Madjar K, *New ChatAI chatbot generates traditional Chinese text*, Taipei Times, 21 February 2023, available at https://www.taipeitimes.com/News/front/archives/2023/02/21/2003794753.
43 *AI and Influence Operations: The Threat of Manipulation at Scale*, Conducttr Blog, 14 January 2023, available at https://blog.conducttr.com/ai-and-influence-operations-the-threat-of-manipulation-at-scale.
44 Ghaffary S, *The makers of ChatGPT just released a new AI that can build websites, among other things*, Vox, 15 March 2023, available at https://www.vox.com/2023/3/15/23640640/gpt-4-chatgpt-openai-generative-ai.
45 Woolley C & Howard P (Ed), *Computational Propaganda Political Parties, Politicians, and Political Manipulation on Social Media*, Oxford Studies in Digital Politics, available at https://global.oup.com/academic/product/computational-propaganda-9780190931414?lang=en&cc=us.

OPSEC

In Chapter 1, I explained the various components and their definitions within IO. One of the least covered parts in the book so far has been OPSEC – Operational Security. That is not to say it is unimportant, far from it. As the conflict in Ukraine shows, OPSEC is vital and yet incredibly difficult to achieve, especially considering the speed new technology is evolving and the fact that people are increasingly wedded to their mobile phones and minded to share data. The ability to locate and track individuals, as well as gaining all kinds of private information from their calls and data usage via mobile devices, will be a growing issue over the next few years and will present problems not just for deployed personnel in areas of conflict, but also for families left at home. This is not new; at the height of the Afghanistan conflict Taliban supporters were alleged to be making 'hate calls' to the UK homes and relatives of soldiers serving in Afghanistan. "Senior commanders believe they get the numbers either by monitoring troops' mobile phone calls or from staff at Afghan phone companies".[46]

Strangely the issue came to more widespread public prominence in 2018 when mainstream media revealed that the fitness tracking app Strava had been giving away the location of secret US army bases.[47] Very popular with athletes of all abilities but particularly those in the military, Strava tracks precise locations and potentially makes the data available to anyone. So it was in 2018 when Strava updated its global heat map of user activity showing walking, running and cycle routes, a student at the Australian National University began posting on Twitter a series of images showing Strava user activities in US military forward operating bases in Afghanistan, Turkish military patrols in Syria, and even a possible guard patrol in the Russian operating area of Syria.[48] The story has been a regular in the British media, in particular, with allegations that the identities and running routes of UK Special Forces were visible.[49] During operations in Afghanistan, I recall that we handed in our mobile phones lest we became traceable.

---

[46] Hickley M, *Taliban tapping British troops' mobiles to taunt soldiers' families*, The Daily Mail, 22 August 2007, available at https://www.dailymail.co.uk/news/article-476959/Taliban-tapping-British-troops-mobiles-taunt-soldiers-families.html.
[47] Hern A, *Fitness tracking app Strava gives away location of secret US army bases*, The Guardian, 28 January 2018, available at https://www.theguardian.com/world/2018/jan/28/fitness-tracking-app-gives-away-location-of-secret-us-army-bases.
[48] Hsu J, *The Strava Heat Map and the End of Secrets*, Wired, 29 January 2018, available at https://www.wired.com/story/strava-heat-map-military-bases-fitness-trackers-privacy/.
[49] Somper J, *Leaking To Vlad. Special Forces troops leak secret locations to Putin on fitness app Strava*, The Sun, 5 September 2022, available at https://www.thesun.co.uk/news/19706949/special-forces-leak-secret-locations-putin-fitness-app/.

But it seems that such discipline has been missing amongst Russian troops in the conflict in Ukraine. The significant number of Russian casualties incurred during the 2023 new year missile attack in the eastern town of Makiivka resulted in Russian General, Sergei Sevryukov, telling the media that his troops phone signals had allowed Kyiv's forces to "determine the coordinates of the location of military personnel" and launch a strike.[50]

Despite the dangers of being targeted, the allure of connectivity to the internet and loved ones at home becomes stronger during the long periods of boredom that characterises all conflict. Researchers interviewed soldiers on the frontline in the Donbas in 2017 and were told that: "Sitting out there in the dugouts, trenches and bunkers for days and even weeks with nothing to do, people start going out of their heads. You need something to take your mind off of things."[51] As well as the immediate danger of being located and targeted, the use of mobile devices has opened up a whole host of other dangers from citizen investigators such as Bellingcat and Anonymous.

Space

There is a fascinating, and sobering video on YouTube entitled 'What if every satellite disappeared'.[52] As the name suggests it imagines a world where satellites have all failed or disappeared. Quickly global transport systems, reliant upon global positioning satellite systems, fail. Global production machinery, much of it reliant upon satellite time stamps to function, ceases. Unable to trade, the world's economy largely shuts down: credit cards can't be read, stock markets crash. Global logistics and supply chains fragment and break and over time, countries declare states of emergency and the world's militaries step in to prevent anarchy and violence. Even in the best-case scenario our civilisation gets set back by decades; the worst case scenario does not bare thinking about.

Traditionally through history there have been three domains of military operations – Sea, Air and Land. Today, however, the US Army

---

[50] Dana & Koslowska, *Sitting ducks? Russian military flaws seen in troop deaths*, ABC News, 4 January 2023, available at https://abcnews.go.com/International/wireStory/russia-phone-allowed-ukraine-target-troops-96169846.

[51] Devine K, *Ukraine war: Mobile networks being weaponised to target troops on both sides of conflict*, Sky TV News, 4 January 2023, available at https://news.sky.com/story/ukraine-war-mobile-networks-being-weaponised-to-target-troops-on-both-sides-of-conflict-12577595.

[52] What if every satellite suddenly disappeared? – Moriba Jah, available at https://www.youtube.com/watch?v=jVzbs81bDy0.

categorises six. Added to Sea, Air and Land, is Sub-Surface, Cyber and Space. So too in the UK, where Space Command was formally established in 2021.[53] The US equivalent, actually set up in 1985 and revitalised in 2019, exists as its deputy director of operations, General Richard J. Zellmann, explained, because "Freedom to access space and operate within the domain has become integral to the American modern way of life".[54]

We have only to look at the Russian war on Ukraine to see why this has become so important. In advance of Russia's invasion of Ukraine in 2022, the US Satellite provider Viasat was attacked and taken down. This was the system used by the Ukrainian Army and was a huge blow to the defensive operation. On 26th February, two days after the cyber-attack, Ukraine's Minister of Digital Transformation sent a Tweet to Elon Musk asking him to provide his army with access to the Starlink system. Within just a few hours, Musk replied: "now active" and within days the dishes needed to access it begin arriving in Ukraine – becoming a vital component in Ukraine's defence.[55] Despite Russia's best efforts, it has not been able to disable Ukraine's connectivity.

China, too, is very interested in Space. Although it only orbited its first satellite in 1970, it has rapidly developed its interests and one of the triggers for that was NATO's ability to defeat Belgrade's air defence systems and pin point target strikes through the use of 86 satellites, providing an almost continuous stream of real time data on Serbian positions.[56] Like the artic regions, Space is likely to become a heavily contested environment and it seems inevitable that both technical and cognitive information superiority of and in Space will form a vibrant area of future IO.

Pace of development

Every day new technical developments and programmes that could have application for the world of IO are announced. For example, 'The Follower' is new software that uses AI to scan CCTV footage to cross-match with

---

53 UK Space Command, Royal Air Force website, available at https://www.raf.mod.uk/what-we-do/uk-space-command/.
54 https://www.spacecom.mil.
55 Campbell M, *Elon Musk's star power is game-changer for Ukraine*, Sunday Times, 15 January 2023.
56 Cheng D, *China's Military Role in Space*, Strategic Studies Quarterly, Spring 2012, available at *https://www.airuniversity.af.edu/Portals/10/SSQ/documents/Volume-06_Issue-1/Cheng.pdf*.

Instagram influencer's photographs and then track them.[57,58] The slightly scary creator of it incentivises its sale with the offer to buy a clock that shows how much percentage of your life is completed based on your life expectancy![59] The huge US data company Palantir have produced a tool called MetaConstellation that allows users to see what commercial data is available for any specific geographic location in the world. And this is not just ethereal technical research; it has real life applications in conflict. In late 2022, The Washington Post reported that every Ukrainian Army Battalion now travels with its own software developer.[60] Napoleon once declared that an army marches on its stomach. Tomorrow an army is more likely to march with their Macs and Memes. Or as a friend wryly observed, armies no longer march on their stomachs, now they must march on the information superhighway! Are we ready?

In March 2023, the UK's Financial Times newspaper reported that the UK MoD had invited six fiction writers to craft short stories on what future warfare may look like. As the newspaper commented, "the stories are certainly creative and provocative".[61] But, as entertaining as they were, they were more descriptive than prescriptive and that in focussing on remote future scenarios – so called Black Swans, the MoD was likely missing what was right in front of its eyes, an 'ugly grey rhino' event. Indeed as the refreshed UK's Integrated Defence Review wrote: "In the context of Ukraine, technology, digital and information warfare have helped to hold back a larger aggressor"[62] and it has pledged to: "develop our broader deterrence and defence toolkit, including IO and offensive cyber tools".[63] IO is the ugly grey rhino, not a black swan and warm words and intent are not enough.

---

57 https://driesdepoorter.be/thefollower/.
58 Wallace S, *Face Recognition Tech Gets Girl Scout Mom Booted From Rockettes Show — Due to Where She Works*, NBC News, 20 December 2022, available at https://www.nbcnewyork.com/investigations/face-recognition-tech-gets-girl-scout-mom-booted-from-rockettes-show-due-to-her-employer/4004677/.
59 https://driesdepoorter.be/thefollower/.
60 Ignatius D, *How the algorithm tipped the balance in Ukraine*, The Washington Post, 19 December 2022, available at https://www.washingtonpost.com/opinions/2022/12/19/palantir-algorithm-data-ukraine-war/.
61 Thornhill J, *Fictional intelligence' can blind us to real-world dangers*, The Financial Times, 9 March 2023, available at https://www.ft.com/content/66bc46a2-a1f6-496b-a467-25751d6e0f8e.
62 The UK Government, *Integrated Review Refresh 2023, Responding to a more contested and volatile world*, 13 March 2023, available at https://www.gov.uk/government/publications/integrated-review-refresh-2023-responding-to-a-more-contested-and-volatile-world.
63 Ibid. 12.

The Law

Every member of the Armed Forces, at least in the global west, has the International Humanitarian Law (IHL), also known as the Law of Armed Conflict (LOAC), drummed into them when they go through basic training. The four fundamental principles of LOAC, military necessity, humanity, distinction, and proportionality, should be well understood. There are of course regrettable lapses even in the West (the case of Royal Marine Sergeant Blackman covered in previous paragraphs an obvious example) but throughout my thirty-five years of military service, and three conflicts, I have never personally experienced anything but the closest adherence to those fundamental principles. In the UK every service person knows that "in every international armed conflict, members of the armed forces, including accompanying civilians, are subject to the law of armed conflict".[64]

And yet as it stands, the Joint Service Publication from which these quotes were taken, JSP383, carries not one single mention of IO. This is clearly a problem. As Thomas Wingfield, former US Deputy Assistant Secretary of Defense for Cyber Policy has written "future wars will feature IO with novel weapons, techniques, and targets. Such IO and cyber attacks will raise unprecedented legal questions to discriminate the lawfully compliant from the negligent, reckless, or intentionally maleficent".[65] Is an Information Operation a 'use of force' that is prohibited under international law? What is necessary force, in IO, to achieve the operational task? Do IO involving cyber-attack satisfy the criteria that care should be taken to avoid damage to civilian objects? IO pose many legal challenges and as a result serve to reinforce the nervousness that military commanders have over their use.

The Oxford Institute for Ethics, Law and Armed Conflict has done much work on this already and in their Statement on International Law, Protections for the Regulation of IO, has declared that "International law applies to all conduct carried out through information and communications technologies, including IO and activities".[66] Well, maybe,

---

64 UK Government, *JSP 383 The Joint Service Manual of the Law of Armed Conflict* available at https://assets.publishing.service.gov.uk/government/uploads/system/uploads/attachment_data/file/27874/JSP3832004Edition.pdf.
65 Wingfield T, *International Law and IO*, National Defence University, available at https://ndupress.ndu.edu/Portals/68/Documents/Books/CTBSP-Exports/Cyberpower/Cyberpower-I-Chap-22.pdf?ver=2017-06-16-115055.
66 Available at https://www.elac.ox.ac.uk/the-oxford-process/the-statements-overview/the-oxford-statement-on-the-regulation-of-information-operations-and-activities/.

but until it is codified by military doctrine Commanders are likely to remain confused at what they can and cannot do. It is not just LOAC that they need to consider. In the UK the Regulatory Investigate Powers Act (RIPA) also applies. RIPA regulates "the interception of communications, the acquisition and disclosure of data relating to communications, the carrying out of surveillance, the use of covert human intelligence sources and the acquisition of the means by which electronic data protected by encryption or passwords may be decrypted or accessed".[67] Another piece of UK legislation that impacts IO is GPR and, post Brexit, the UK's version – The Data Protection Act of 2018. Both essentially enshrine the same responsibilities for using personal data and define rules ('data protection principles') that must be adhered to. For example, they must make sure the information is: used fairly, lawfully, and transparently. This latter point, for example, is likely problematic if automated MoD personas are messaging known social media accounts. Is that transparent?

A final legal issue is Freedom of Information (FOI). At the moment anyone can request data about MoD IO. There are of course security caveats applied to all FOI requests but nevertheless many academics have used the FOI system to publicise operations and programmes. For example, Dr Briant, who we met in earlier chapters, has been a vociferous user of FOIs to research UK and NATO IO. The problem is not so much the request – I am a supporter of well-regulated FOIs as a key component in any democratic society – but how the subsequently derived information, which might come from multiple but related FOI requests, is pieced together and presented by the instigator and what use it may be to an adversary. Again, in Chapter 11, I discussed how information is presented through 'priming' and 'framing'; it is very easy for malevolent actors to present IO in a sinister way and entice all kind of comment about anti-democratic values. Some elements of government are exempt from FOI, the intelligence services and the Special Forces are obvious examples and I would certainly like to see greater consideration of IO for FOI exemption.

We may also need in the future to consider the export licensing of IO equipment and methodologies. When I ran the trial of the BD methodology, I and my colleagues were worried that it could be exported to unfriendly nations; we investigated with the then Department of Trade if it could be placed under export control. That never materialised, it was explained

---

[67] UK Government, Regulation of Investigatory Powers Act, 2000, available at https://www.legislation.gov.uk/ukpga/2000/23/introduction.

that export licences were only applied to physical items of equipment – weapons, radars etc – and not methodologies. As IO usage and technology expands and develop this may require review.

When I commanded UK PsyOps in Helmand, we were very aware of the risk tolerances of the various Commanders under whom we operated. Some were content to let PsyOps work largely unsupervised. Others insisted that every element of PsyOps be scrutinised by the Political Advisor (PolAd) and the Legal Advisor (Legad) on the staff. The difference in views over what was and was not legal was often surprising and at times a barrier to undertaking operations. It is the Legads in 77the brigade, in MSE, in 4th POG, in NATO *et al* that are likely to have to offer advice on future IO and it's not clear to me, presently, that a sufficiently robust corpus of law and direction currently exists. Whilst Russia and Iran would not be troubled operating outside the law (and often do), that is not a direction the UK and other western armed forces must ever take, and the legal elements of IO need very urgent attention.

13 Conclusions

The most notable change in warfare over the last fifty years is information and over the last 231 pages I have sought to show what IO really is, not what certain sections of social media and academia would have you believe, and in particular show you how its use, which is very carefully prescribed, often very conservative, in NATO armed Forces (if not in Russia and China), will increase and to be successful will need to become far smarter in understanding the audiences that it is deployed within. I have shown you the nature of fake news and how that will become a very real threat to our democracies as AI becomes better and cheaper. And I have shown that there are plenty of conspiracists who play on the public's lack of knowledge of IO and, in particular, PsyOps to demonise, denigrate and undermine.

Throughout the book I have only occasionally mentioned cyber. This is deliberate and for two reasons. Firstly, by their very nature, cyber operations are extremely sensitive and there is little that I can safely place in the public domain or would even wish to. Secondly, no matter how good the technology, ultimately IO are about attempting to change human behaviour and it is on that I have chosen to major, because whilst we are very good at cyber we are less good at changing behaviours. What I can say is that I have witnessed the development of an impressive, rapid and innovative technical cyber development programme over the past 20 years, and it fills me with hope that we might bring the same innovation and thoughtfulness to our broader IO. And it has worked because people who know what that were doing were empowered; IO is full of good, well-meaning, people. But it is not full of people who can draw on a deep well of experience of knowledge. The fact that this book took over a year to clear for publication is indicative of this.The film, *Miracle on the Hudson*, tells the story of Captain 'Sully' Sullenberger, the pilot of American Airlines Flight 1549 who successfully landed his stricken aircraft on New York's Hudson River.

The film shows the subsequent investigation, including the various hearings, in which he is criticised for not landing the aircraft at one of the available New York airports – which various computer simulations had shown to be possible. In the film Sully asks: "Can we get serious now? We've all heard about the computer simulations, and I can't believe we are watching actual sims, but I don't quite believe we still have not taken into account. the human factor".[1] In my view the very same point applies to IO.

In the years after 9/11, IO and its practitioners became entranced by the ideas of advertising and marketing companies. Commercial companies made millions of dollars from advertising and marketing-based communication programmes in Iraq, Afghanistan and in the wider global war on terror. The USA Today correspondent Tom Vanden Brook has written on US IO for some years. His many articles chart the huge expense (for example, "spending on IO reached upwards of $580 million in 2009, a number which may be staggering to professional public diplomacy practitioners"[2]) and the lack of obvious success. Although he is unpopular with many in the US IO community this is not to say that his opinions do not have merit. In private many colleagues, from across the international landscape, look back on IO efforts in Iraq and Afghanistan and ask similar questions. Immediately after 9/11 the clamour to 'do something' was huge (and entirely understandable) and partly because the West is an advertising and marketing saturated society many senior decision makers happily grasped at the straws offered by big business.

That is not to say everyone followed the herd; in the US Professor Montgomery McFate, for example, managed to find funding for a small number of Human Terrain Teams (HTS) – although not enough to turn the tide in any meaningful way in Iraq or Afghanistan. Had there been more, and had their recommendations been listened to, arguably we may well have seen different outcomes. In the UK a tiny number of military officers were selected for language and cultural training (the so called CULADs) and finally got out on the ground but as Captain Mike Martin (a CULAD in Helmand, Afghanistan) describes in his excellent 2014 book, they were only really able to tell the command how badly things were going.[3] Like their US counterparts they were few, far apart and seen as something on the periphery of military operations. And no one liked bad news.

---

1 Akillheals, *Sully Scene, Can we get serious now*? YouTube, available at https://www.youtube.com/watch?v=N1fVL4AQEW8.
2 Wallin M, *The Dollars and Dimes of Hearts and Minds*, American Security project March 2013, available at https://www.americansecurityproject.org/the-dollars-and-dimes-of-hearts-and-minds/.
3 Martin M, *An Intimate War: An Oral History of the Helmand Conflict*, Hurst & Co, 2014.

Through the battles with ISIS the same problems prevailed; IO was never really regarded as a mainstream activity. In August 2013, for example, the UK Parliament debated, and subsequently decided against, military action in Syria following President Assad's use of chemical weapons.[4] My colleagues in the kinetic targeting office of Military Strategic Effects had spent some weeks working across government to come up with different target packages – in other words, planning where the bombs would land. I and my colleagues in the IO office had instead looked at how we could punish Assad with a non-lethal IO programme – we felt we could trigger a crumbling of Assad's power base through highly targeted IO. Yet our plans were never presented to the MoD's senior leadership.

Today Russia, whom many once hoped might be an international partner and not an enemy, is once more the UK's top security threat.[5] In January 2023 the Chairman of the NATO Military Committee warned of all-out war with Russia in the next two decades.[6] It's easy to see why. Whilst Russia worries about NATO troops being on its western borders, Russia has its own forces across parts of Eastern Europe – in Ukraine, Moldova, Georgia – as well as in Venezuela,[7] in Vietnam and, in the future quite possibly across Africa.[8] As we have seen, Putin is unafraid to unleash IO, to distort and deceive, and he appears to have significant assets for that task, military and irregular. Yet in the West, despite the Ukraine War, despite all that we know about Putin's propaganda, we seem still to be without key enablers in our IO armoury, notably the social scientists and behavioural psychologists. To deliberately re-quote Tom Hank's Captain Sullenberger "I don't quite believe we still have not taken into account the human factor".

As I write this chapter in 2024, riots and civil disorder continue across Iran – a country with an incredibly well-developed internal security and

---

4 Gorjestani K, *UK parliament votes against military action in Syria,* France24 TV, 30 August 2013, available at https://www.france24.com/en/20130830-british-parliament-rejects-military-action-syria-cameron-assad.
5 Fisher L and Rathbone J, *Russia remains UK's top security threat warns defence intelligence chief* The Financial Times 30 May 2023, available at https://www.ft.com/content/57216d44-924c-409f-912b-fa87d52e0021.
6 Kalsi G, *Nato braced for all-out war with Russia in the next 20 years,* The daily Mail, 18 January 2024, available at https://www.dailymail.co.uk/news/article-12981021/Nato-braced-war-Russia-20-years.html.
7 Ward A, *Why Russia just sent troops to Venezuela,* VOX, 27 March 2019, available at https://www.vox.com/2019/3/27/18283807/venezuela-russia-troops-trump-maduro-guaido.
8 Bledsoe V, *How Many Military Bases Does Russia Have Overseas,* The Soldier's Project, 2 April 2023, available at https://www.thesoldiersproject.org/how-many-military-bases-does-russia-have/#11_In_Africa.

intelligence service. But for all their agents and networks of informers, they appear once again surprised by the widespread actions of so many of their population and, as we saw in previous chapters, Russia completely misjudged the willingness of the Ukraine population to resist its invasion in 2022 – including the very Russian speaking Ukrainians who allegedly Putin's war had been to protect.⁹

There is a constant line back through recent history of nations with very well-developed intelligence systems being completely caught off balance by the actions of populations; the Arab Spring (which contrary to some commentary was never caused by social media) caught many regimes off guard, not least in Egypt where the presence of over 200,000 people in Tahir Square proved an impossible problem for the regime to resolve. In Zimbabwe, long term President, Robert Mugabe, woke up one day to find he was no longer the doyen of his Zanu PF party and that he had been expelled – again a nation with well-equipped internal surveillance and intelligence structures unable to track what was happening within their own population. In the chapter on Iraq, I noted how the majority of the UK's problems in southern Iraq were caused not by the Sunni but instead by the Mahdi Army – something we knew very little about – and its hither too unknown leader, Muqtada al Sada, who had returned from exile in Iran. In an article for *The Sunday Times* in early 2024 the eminent British historian Sir Laurence Friedman considered why the much larger and numerically superior nation of Russia was having so much difficulty in Ukraine post the February 2022 invasion. He wrote, "The main reason Putin failed was that he had a caricature view of Ukraine that assumed that it was an artificial state with a weak, illegitimate government and would barely resist. In this he fell into the trap of totally underestimating his opponent".¹⁰

Unless we develop comprehensive Population Intelligence capabilities built not on social media scraping but on the social sciences, we are likely to continue to be caught by surprise. Our trials of the BD methodology demonstrated that doesn't have to be the case. Properly conducted field research could be undertaken safely and ethically and used to inform western governments of what was actually going on inside populations

---

9 Gormezano D, *Russian speakers reject the 'language of the enemy' by learning Ukrainian*, France24, 30th May 2022, available at https://www.france24.com/en/europe/20220530-russian-speakers-in-ukraine-reject-the-language-of-the-enemy-by-learning-ukrainian.
10 Friedman L, *What the Russian-Ukraine War tells us about future Conflict*, Sunday Times magazine 4 February 2024, available at https://www.thetimes.co.uk/article/the-future-of-war-3dd3pg86r .

and nations of interest so they could be ready with appropriate response options. This had been a very long journey fraught with both internal and external resistance but finally the UK MoD, the US DOD and NATO were starting to understand the utility of applied social science research and the need to understand the human factor. They began to realise that if you held that data you then knew what to say, who to say it to and how to convey it. Combining Population Intelligence with advanced social media campaigns and cyber operations would allow precision guided information, as opposed to indiscriminate messaging. And if you properly collected the data, you would also be able to assess your progress. No longer would IO campaigns be 'measured' on their performance (how many 'likes', how many 'retweets', how many comments) but instead on the actual effectiveness of the transmitted message on changing behaviours. Indeed, in this latter regard our assessment of the effectiveness of IO has not really progressed much from the Soviets in Afghanistan in the 1980s. "Activities were evaluated not according to the actual results achieved but according to the number of measures carried out",[11] wrote Soviet General Serookiy in his study of the Soviet Psychological Operations campaign during their Afghan war.

What is so enormously frustrating is that we had that capability in 2014. It was then that I and my colleague, a Colonel in the Intelligence Corps, managed to secure the necessary funding to demonstrate the BD audience analysis in a real scenario and have it formally assessed by the MoD's client-side advisor, DSTL (Defence Science and Technology Laboratory). I still have photos of the presentation of the findings of the audience analysis to over 100 representatives from across all government departments, from the US, Australia, Canada, and NATO in the pillared hall of the MoD. I remember the sense of excitement as people came over and congratulated us – they all saw the value in Population Intelligence; they saw it was something we didn't have, and it was something we needed. Within weeks of the trial finishing, we were asked to put it to use on a real operation. It was entirely successful. And yet, the lessons were never institutionalised and as those that were involved moved away to other roles the corporate knowledge died. The 'Thinkdefence' website[12] is a blog that promotes the idea of people having a sensible conversation about UK defence issues. In February 2024 they tweeted "the MoD's corporate memory is appalling, records are not

---

11 Serookiy Y Y, *Psychological Information Warfare Lessons of Afghanistan* Military Thought 2004-03 No 001.
12 www.thinkdefence.co.uk.

kept, the army spends more time learning about Waterloo than it does about failed programmes".[13] And successful ones too! Yet today there are active programmes of research, supported by the British Government and costing up to one million pounds, into testing and evaluating different audience analysis methodologies, absolutely repeating that undertaken in 2014.[14] This is not a good use of public money and has happened because today no one across the wider defence community has any institutional memory of what had gone before, at least in IO.

What Do We Need To Do?

On current trajectory, the UK and the West do not appear to be heading down the precision information weapon path, instead the strategy appears to be the same anonymised mass social media messaging as the US have used. This is disappointing. Not only are we seemingly following the same model as adversaries such as Russia – and we have higher values and standards than them – but any Psychologist will tell you those campaigns just don't work very well. As the Washington Post reported on the US' anonymised social media efforts: "the clandestine activity did not have much impact".[15] It noted that the vast majority of posted tweets received "no more than a handful of likes or retweets and only 19% had more than 1000 followers". It went on to say that the two most followed accounts were "overt accounts that publicly declared a connection to the US Military".[16]

Why would it have much impact (and what impact was expected anyway)? We are so bombarded with information and adverts these days – the average person sees between 6,000-10,000 advertisements a day[17] – that we are becoming immune to them. Indeed, paradoxically the PR companies are very aware of this. The inadequacy of the old marketing model, and the advertising industry it supports, is increasingly turning to science to provide solutions amid the transformation wrought by the

---

13 @thinkdefence 19 Feb 2024 9.27pm available at https://twitter.com/thinkdefence/status/1759691173176701315.
14 *Up to £1 million available to develop technologies that help understand audiences*, UK Government, 21 March 2023, available at https://www.gov.uk/government/news/up-to-1-million-available-to-develop-technologies-that-help-understand-audiences.
15 Nakashima E, *Pentagon opens sweeping review of clandestine psychological operations*, The Washington Post, 19 September 2022, available at https://www.washingtonpost.com/national-security/2022/09/19/pentagon-psychological-operations-facebook-twitter/.
16 Ibid.
17 Carr S, *How Many Ads Do We See A Day In 2023?*, Luni, 15 February 2021, available at https://lunio.ai/blog/strategy/how-many-ads-do-we-see-a-day/.

communications revolution. And as we saw in an earlier chapter, some social media companies are finally clamping down on fake accounts and insisting that verified accounts are paid for. That trend, together with undoubted greater interest in regulation by countries in which social media channels operate (and we see the first attempts at this, in the EU, with the Digital Services Act[18]), is only going to continue and it does not bode well for mass disguised personas.

During the World War II, the UK adopted a whole of government strategy to information focussed on overcoming the existential threat of Nazi Germany. Arguably we face an existential threat today to our societies and we again need a whole of government effort because the military does not and must not operate in a vacuum. We can therefore start by addressing some of the deficits in our national communications.

First, we need a long-term communication strategy rather than the short termism that seems to permeate all of our information activities. In this regard we could usefully learn from Russia and China who see far 'bigger pictures' than the West. Kier Giles, long term Russia analyst, likens the West's IO to FM radio frequency modulations – up and down over time – whereas in Russia and China he likens their IO to a radio carrier wave – strong, constant, and enduring.

Second, we need to stop the cuts to the BBC World Service, and we need to resource and expand BBC Monitoring back to its previous levels. To have ceased the broadcasting of BBC Arabic and various BBC African language channels is hard to understand. If we are to prepare the information space in the way I suggested earlier, so that it is our adversaries that have to defeat our well-established narratives, then the BBC World Service is vital. And so too BBC Monitoring; I, along with many others, wrote to Parliament when its potential disestablishment was announced. BBC Monitoring provided an exceptional insight into events on the ground and proved invaluable in so many previous operations. We are poorer for its reduced presence.

Third, some humility. Policy makers need to step outside their own echo chambers, be humble and acknowledge that the West does not necessarily have compelling political messages for large chunks of the world. Our democratic values do not carry automatic recognition or agreement globally. As The Economist wrote, "at least 4 billion people, or more than half of the world's population, live in over 100 countries that do not want to

---

18 *The Digital Services Act package*, European Commission, available at https://digital-strategy.ec.europa.eu/en/policies/digital-services-act-package.

pick sides [over Ukraine]".[19] As I showed in Chapter 7, nations such as the United Arab Emirates are very happy to continue their relationship with, for example Russia, and are not convinced by what they regard as Western double standards – the lack of visible support for Palestinians in contrast to the West's billion-dollar support for Ukraine is an obvious example. We ought not to forget too that autocratic regimes are more likely to rally to nations such as Russia than to democracies. We continue to see the world as we wish it to be, not as it is.

Fourth, our overt presence in the information environment must be strengthened and, importantly, synchronised. It would not hurt to try to foster a unique UK / British identity, separate from whatever political party may be in power at the time. 'Britishness' – by which I mean the totality of our nation spanning, for example, the arts, history, the Royal Family, culture, business, governments, LGBTQ+ and ethnic minorities needs to be managed and messaged. Indeed, instead of stoking endless internal culture wars, politicians might usefully think how our culture can be promoted externally. In the mid 1990s the idea of 'Cool Britannia' emerged, was much criticised and eventually faded. We need an enduring strategy that is separate from political identity (and thus can weather political storms such as Brexit) that is appealing and importantly provides a very discernible contrast with the Russian and Chinese offers.

Fifth, and although not a fan of covert messaging, I recognise volume and repetition can make a difference, that it has a place, but it must be subordinated to, and support, the overt presence. Covert messaging should only amplify and not lead discussions. We are not Russia or China and we should not resort to the same techniques of Bot Farms and Fake News. If we are to use fake personas, they must solely be as signposts to overt messaging. We might usefully consider the words of Matthew Syed who offered The Times his view of much of what is online today: "people dont believe the bullshit circulating online. They are breathlessly amplifying this nonsense precisely because it is fantastical, escapist, felonious; it is a way of switching off from real life".[20]

Sixth, we must start publicly calling out the 'useful idiots'. Finally, the UK is following the US lead and the Foreign Influence Registration Scheme

---

19 *Can the West win over the rest?*, The Economist, 14 April 2023, available a http://www.economist.com/leaders/2023/04/13/can-the-west-win-over-the-rest.
20 Syed M, *Her Kindness to my son points up the cruelty of algrorithms and fools*, The Times, 24 March 2024.

is passing through Parliament;[21] anyone that is lobbying or representing a foreign power must be registered and if not, they need to be held to public account. It cannot be left to social media companies alone to label individuals as representatives of 'Foreign State Media'. Former infantry officer and now editor of UKlandpower.com, defence consultant Nicholas Drummond recently wrote that: "The extent to which Russia and Iran are using information warfare to promote their warped agendas should alarm us all. They are funding academics, politicians, journalists, and other people in influential positions… we repealed our sedition laws because of their effect on free speech, but our treason laws are too narrow to catch these people. The power of social media demands new laws that make people accountable for the hateful and destructive things they say online".[22] We must always value freedom of speech but with that freedom must also come obligations.

Seventh, national resilience. We need to be building resilience amongst our population as well as skills in critical thinking; this seems best placed in the educational curriculum and delivery should begin at a very early age. The development of such skills has already been the subject of much research with some concluding that "interventions that fostered an analytical mindset or taught critical thinking skills were found to be the most effective in terms of changing conspiracy beliefs".[23]

And eight, our messaging is strengthened when it comes from others – we are very poor at identifying international influencers who might have reach and depth into audiences that ordinarily we would have no chance of addressing. One person in the right conversation at the right time is worth whole armies of Twitter bots. This latter point leads me to my overarching and now oft repeated point. We need to combine advanced Population Intelligence, which tells us who to say what to and how, with social media, mainstream media and word of mouth campaigns and, if necessary, cyber operations to create precision guided information weapons, as opposed to indiscriminate messaging 'noise'.

As this book is principally about the military let me return to the Armed Forces. What can they do to improve?

---

21 *National Security Bill: factsheets*, UK Government briefing Paper, 6 June 2022, available at https://www.gov.uk/government/publications/national-security-bill-factsheets/foreign-influence-registration-scheme-firs-national-security-bill-factsheet.
22 @nicholasdrummond 20 Feb 2024 2.57pm, available at https://twitter.com/nicholadrummond/status/1759955427712696380.
23 O'Mahony C, Brassil M, Murphy G, Linehan C, *The efficacy of interventions in reducing belief in conspiracy theories: A systematic review*, available at https://journals.plos.org/plosone/article?id=10.1371/journal.pone.0280902.

First, it seems to me that there needs to be bespoke senior leadership at least at the two-star Officer or Civil Servant level. The US has done this with the creation of a Principal IO Advisor to the Under Secretary of Defense for Policy. In the UK we do need to corral the various organisations that exist across UK defence and to bring the right people together to build an effective capability. Such a post might help generate a more nuanced understanding across the MoD of what can and cannot be achieved. It simply cannot be that Government Ministers and Senior Officers look at what Russia has done in western democracies and ask, 'so where is our bot farm'?

Second, we need fit for purpose organisations. 77th Brigade comes solely under Army control and was a smashing together of the various units its consumed such as 15(UK) PsyOps, and the UK's Civil Military Affairs team. The need for the type of capability that 77th is designed for is not restricted to the Army. I know that this has been debated at Army Board level in the past, but it seems obvious to me that 77th Brigade should be part of the UK's joint capability vested in Strategic Command. And were that to happen, 77th Brigade staff, and particularly leadership, would not be picked solely from the ranks of the Army but from across the wider defence community. Some of defence's core and enduring IO experience sits with the nearly 200 men and women of the Royal Naval Reserve Information Operations specialisation who spend their entire careers in IO and yet will never get to command 77th Brigade. It is anathema that every Commanding Officer of 77th Brigade to date has been a 'teeth arms' officer, as if being in the infantry or cavalry prepared you for the intricacies of information warfare. Having worked closely with three of them – all good people – they were honest enough to admit to me that they took on the role knowing absolutely nothing about the subject and having to learn on the job. As they themselves told me, they felt that their roles owed much to internal presentation in the Army, because without a 'going places' senior teeth arms Brigadier at its helm, 77th Brigade would struggle in the face of budgetary fights between 'real' teeth arms such as tank regiments and the infantry.

Third, we need qualified and experienced people, not just enthusiastic amateurs, to stand up a professional IO cadre of both regular and reserve tri-service personnel. There will be huge resistance to this because it will strip assets from other places in defence but until we do, and provide incumbents with a regular career stream, we will continue to have significant and sustained difficulties.²⁴ One innovative approach might be to create a joint US/UK cadre of personnel. In Chapter 3, I discussed the US and its huge resources. Yet for all its mass and money, the US has not dissimilar problems to the UK. As a professional PSYOP Officer in the US Army you will never be promoted to General Officer rank – the furthest you can expect to progress, and there are very few, is full Colonel. Yet in contrast, if you are a Public Affairs Officer in the US military, you can be promoted to a starred rank – Generals and Admirals. Indeed former US Secretary of State for Defense under President Trump was Admiral Ken Braithwaite – a former US Navy public affairs officer! The message this sends is 'we care more about what the papers say than fighting an adversary in the information domain'. If the US is not prepared to elevate their most highly experienced and qualified IO officers to the senior most levels, what chance have smaller nations such as the UK? This has not gone unnoticed. Lieutenant General, Robert Elder (USAF, retired), wrote: "Because of the US military's traditional reliance on attrition warfare, few US military leaders are experienced with the use of IO to outmanoeuvre an adversary as a means to victory. However, as a means to combat the US asymmetric advantage in physical power, many US competitors have turned to information as a means to outmanoeuvre the United States".²⁵ A broadening of opportunities for both the US and the UK could be a smart move.

Fourth, the military need access to enduring social science support to facilitate a laser like social science focus on behaviours not attitudes. For that we need access to academia. Not sensationalist academics and useful idiots, but sensible academics with appropriate security clearances, willing to help the UK develop its information capabilities. We could perhaps leverage off the US academic research programme with the University of Maryland's ARLIS centre, one of 14 Department of Defense University-Affiliated Research Centers (UARCs) and the only one with a core mission to support the government's security and intelligence communities. ARLIS undertakes research into the social and behavioral sciences, AI, and computing for IO.

---

24 At the moment UK defence's expertise exists within the Reserve Forces and principally, but not exclusively, the Royal Naval Reserve Information Operations branch. Highly trained and very experienced they augment regular units globally but they are reservists and their ability to sustain operational support is not in-exhaustible and constrained by their full time, civilian, employment. Nor will a reservist ever be allowed to command an organisation such as 77X.

25 Elder R, *Information Manoeuvre in Military Operations*, George Mason University, August 2021, available at https://nsiteam.com/social/wp-content/uploads/2021/08/IIJO-Invited-Perspective_Info-Maneuver-in-Mil-Ops_FINAL-2.pdf.

Fifth, and presuming we have the preceding four in place, we need to start a different engagement with industry. Militaries are very reliant upon the private sector to support IO at scale. There is nothing wrong with that and I named some of the many companies that get big contracts in Chapter 11. Yet because of the client's lack of experience (i.e. the UK MoD, NATO and, to a lesser extent, DOD) most of those contracts continue to stipulate contractual requirements in terms of 'performance'. For example, the winning company must be able to distribute 500 videos on line or mass produce 1 million leaflets. There are always contractual stipulations that the products must be 'culturally relevant' or 'cater for specific languages' but these tend to be superficial and the overall contractual framework and the key performance indicators (KPIs) are almost always measures of performance. That is absolutely the wrong metric but until the clients stipulate that the performance must be measured by their effectiveness, which would force industry into undertaking meaningful research of the type BD developed, our information efforts are likely to continue to be lacklustre. And we will not get those smarter KPIs until clients have experience and knowledge and can write good contractual statements of requirement.

Sixth, in professionalising we might usefully follow the example of the US which has established the Information Professionals Association (IPA),[26] an accredited body which aims to professionalise and demystify IO; such an organisation could be helpful in offering appropriate accredited training and certifications and acting as a reputable spokes body when major news stories about IO, Influence and PsyOps break.

Seventh, we might look to second people from other government departments, and maybe other nations, to help. The UK Armed Forces very often second officers to other organisations but very rarely take secondees. Can we learn anything from other UK government departments or from industry? I think we can. Take UK Policing for example. BitCoin blackmailer, Emil Apreda, who threatened to blow up an NHS hospital during the Covid pandemic, was tracked and finally prosecuted using behaviour science research.[27] Police behavioral investigative advisers speak at length in episode 2 the '*Profiling and the Murders of Yvonne Killian and Rachel Nickell*', and how they support police investigations in writer Lynda

---

26 https://information-professionals.org/about/.
27 Dearden L, *Neo-Nazi' jailed for threatening to bomb NHS hospital at height of pandemic*, The Independent, 26 February 2021, available at https://www.independent.co.uk/news/uk/crime/nhs-hospital-bomb-threat-jail-covid-b1808179.html.

La Plante's crime podcast. The body of 6-year-old, Rikki Neave, who was murdered and hidden in woods near his house, was found nearly 30 years later, and his murderer convicted, using behavioral science advice.[28]

There should be no surprise – this is the result of leadership and deliberate policy. When Professor Paul Taylor took over as Chief Scientific Advisor to the Police in 2021, he would find his tenure coinciding with the British Governments intent to: "make the UK a global science superpower by 2030".[29] The study of behaviour is gradually becoming 'core business' across some elements of western governments. The UK government established a Behavioural Insights Team some years ago to design better public policy.[30] The British Security Service, MI5, openly advertises its use of behavioural scientists in operational investigations, on their website.[31] GCHQ – the UK's governments code breaking and listening centre – also uses its website to advertise its requirement for behavioural scientists.[32] The Police have been no different. The MoD is very happy to recognise the importance of behaviour and psychology in so much of its day-to-day work. In July 2022 the MoD issued 'Psychological Safety In MoD Major Projects – Creating the environment for our projects to succeed'.[33] The Royal Navy has issued 'A Guide to Understanding Human Factors & Human Behaviour in Safety Management and Accident Investigation'.[34] And yet in IO the social scientist has been replaced by the IT man; psychologists and social science research by web scraping; and behaviours by attitudes and perceptions.

---

28 *The Boy in the Woods* available at https://www.bbc.co.uk/programmes/m001byjx/episodes/player.
29 Vallance P, *UK's quest to be a global science superpower,* 8 February 2022, available at https://civilservice.blog.gov.uk/2022/02/08/uks-quest-to-be-a-global-science-superpower/.
30 Sanders, Snijders and Hallsworth note in their paper, Behavioural science and policy: where are we now and where are we going?, that the wide-ranging transformation of public policy development that many thought possible has remained absent. One reason given for this is the absence of more 'diagnostic' approaches, including better tools and models, to ensure that behavioural science is not perceived as offering merely technocratic tweaks – that diagnostic approach being the basis of the BD methodology. See https://www.cambridge.org/core/journals/behavioural-public-policy/article/what-are-we-forgetting/677B635D8C51E4201D8598517A4F79C5.
31 Job advert, Specialist roles at Mi5, available at https://www.mi5.gov.uk/careers/opportunities/specialist.
32 Job advert for Applied Behavioural and Social Scientists, available at https://www.gchq-careers.co.uk/jobs/applied-behavioural-and-social-scientists.html.
33 Available to download at https://assets.publishing.service.gov.uk/government/uploads/system/uploads/attachment_data/file/1099704/Psychological_Safety_in_MOD_Major_Programmes_Report.pdf.
34 Available at https://www.era.europa.eu/system/files/2022-10/Navy%20-%20Understanding%20Human%20Factors%20Guide.pdf.

And what of other nations? Germany, the US, Italy, Romania and Poland all have professional IO and/or PsyOps personnel in their armed forces. Military exchange programmes have long seen officers from other nations taking exchange posts to the UK. Why not in IO? Which leads to point eight.

We need to unite NATO IO forces in a tighter group than is currently the case. They may well meet at yearly working groups, but interoperability is largely non-existent and equipment and training largely non-standardised. And there are particular nations that now have significant expertise – Finland (which set up its capabilities in 2014) and Ukraine are obvious examples. NATO school at Oberammergau could play an important role in developing the wider NATO capability as could the NATO Centre of Excellence in Latvia, who we should remember was the recipient of one million Canadian Dollars for advanced BD audience analysis training.

Eight years after designing and managing the UK's largest and most expensive behavioural based audience analysis trial, 11 years after co-writing with Generals Andrew Mackay, Stanley McChrystal and psychologist Dr Lee Rowland of the need to understand behaviours and their drivers in *'Behavioural Conflict'* and 17 years after writing in *'Losing Arab Hearts & Minds'* of our collective failure to understand Arab and muslin behaviour post 9/11, I find myself writing yet another book saying the same. Our collective IO effort has not progressed as meaningfully as it should have done given the wealth of operational lessons we learned, and the money spent. German philosopher Arthur Schopenhauer once said that all truth passes through three stages. First, it is ridiculed. Second, it is violently opposed. Third, it is accepted as being self-evident.[35] We battled for years with BD, through phases 1 and 2, before finally reaching 3. And yet in a few short years we appear to be back to Phase 1. We are still sending messages that 'we' invent to audiences that 'we' think are relevant without any or much understanding of the context in which those audiences exist and with almost no regard to social science, which remains largely absent from the UK, NATO and sadly now the US with the loss of the HTS.

And those operations, in the UK at least, remain overseen by senior officers with almost no background in behavioural change and actioned by junior officers who undertake the work for a short period of time before returning to their 'core' military roles which will get them promoted. This

---

35 Chu M, The 3 stages of truth in life, Huff Post 28 Jul 2016 available at https://www.huffpost.com/entry/the-3-stages-of-truth-in_b_11244204.

seems remarkably close to the claimed Einstein definition of madness: 'Doing the same thing over and over again and expecting different results'. As a very senior US friend of mine, who had returned to IO after an eight-year career break, confided in me: "nothing has changed – we haven't moved on".[36] She is right. We continue to fail to ask the fundamental question that should start any consideration of IO: 'Under what circumstances would an audience do (or not do) something'. If you ask this question everytime you force yourself down a path of understanding rather than just leaping to messages.

Many senior leaders still remain sceptical of IO because it is seen as being of less value than hard military power. Certainly the Cambridge Analytica debacle has not helped at all and there is a perception, as a result of Cambridge Analytica, that audience analysis is somehow tainted. Time and time again I hear that battles cannot be won through IO alone. Probably true, but conflicts can be deterred in the first place by IO – if accompanied by the right audience research. And battles, and the appalling loss of life that war fighting causes, can be reduced by good IO. But until we get serious about it, we have no hope of achieving its potential. And until we invest in the underlying precursor for success, Population Intelligence of the type we tried and tested in BD, not junk scraped from the internet, then these aspirations remain unachievable.

There is no such thing as a Lifetime Achievement Award in IO. But if there was, a worthy recipient would be the originator of the BD methodology – Nigel Oakes. Nigel spent the best part of his life developing an increasingly robust method to understand audiences for use in military IO. As a result of his work, audience analysis and the need to differentiate between attitudes and behaviors has become better understood – certainly amongst the generation who served in Afghanistan and Iraq. But then of course Cambridge Analytica erupted, and the name Oakes became almost heretical to mention. In my long career in IO, the BD methodology was perhaps the most important and useful development I witnessed and the fact that it has been completely forgotten is utterly tragic.

There is an irony that elements of academic, the media and sections of civil society were so outraged by the very idea of understanding audiences to counter fake foreign IO yet at the same time were expressing concern at the extent of Russian influence in British society. As a practitioner of nearly

---

36 Discussion with former senior DOD official and academic at Exercise Phoenix Challenge April 2023.

thirty years' experience there is no question in my mind that some of the most significant human factor advances that I have personally witnessed were ones that Oakes and the Behavioural Dynamics methodology instigated.

As I look forward, I see IO becoming more and more important in everyday life – not just for the military. In banking, in logistic supply chains, in the media, agriculture, in the entertainment industry – all have the potential to be impacted by our adversaries' IO. Indeed, the collapse of US bank SVB in early 2023 was felt by some to have been at least in part caused by social media panic. As The Economist wrote at the time, "the banking turmoil that has sent a handful of American and European lenders to the wall in recent weeks has a new feature. Use of social media and messaging apps, which spread information at lightning pace to an ever-larger group of panickers".[37] That type of panic could be distressingly easy to foster through IO. Most commercial organisations I have come across have plans in place to deal with cyber-attack to their computer systems. But what of an information Operations attack? Shell was hit some years ago when a fake press release was released saying they were suspending offshore drilling of the coast of Nigeria.[38] As we saw in an earlier chapter, BAE Systems were victim of a fake letter from Swedish foreign minister thanking them for demonstrating their missiles for Ukraine. But is it just the big multi-nationals? Consider a small software start up that is about to be bought by a bigger company. There is a period of due diligence and purdah; imagine if some staff member is aggrieved and decides to orchestrate a fake news campaign, maybe using multiple social media counts to allege that the software company CEO is part of a secret paedophile ring. That could completely derail a take-over or an IPO

More and more nations are embracing IO; India, Brazil, Vietnam, Indonesia, Pakistan, and Australia, and New Zealand all have embryonic or developing capabilities and that list is only going to expand. The concern is do we have the ability to understand them, where necessary to combat them and to ultimately prevail in the information environment? Given the tsunami of contrary evidence there is still a widely held view that audience analysis is merely polling. There is a view that domination of social

---

37 *Did social media cause the banking panic?*, The Economist, 30 March 2023, available at https://www.economist.com/finance-and-economics/2023/03/30/did-social-media-cause-the-banking-panic.
38 Reuters, *Hoaxers target Shell with bogus Nigeria news*, 17 May 2020, available at https://www.reuters.com/article/idUSLDE64G28B/.

media is possible and is the answer to everything. Measures of Effect and Measures of Performance are still routinely conflated and misunderstood and key metrics such as Locus of Control and Normative Affiliation remain unknown. It is my hope that this book might just stimulate some introspection and thought. Very deliberately repeating myself, I hope that after reading this book people will understand the importance of asking 'under what circumstances might an audience do or not do something' rather than leaping to messaging. Understanding audiences is the key to success and it is badly lacking.. Regardless this is both the literal and metaphorical end chapter for me – after nearly 30 years in IO it is time to draw to a close.

I end with the words of World War Two British propagandist John Baker White, who concluded his book *The Big Lie*, thus: "After the war I met a General with a fine fighting record. In his best inspection manner, he asked 'And what did you do in the war?' I replied 'Mostly psychological warfare, deception and kindred activities'. His mouth curled visibly with disgust and he stepped back as if contact might pollute him. 'Filthy game' he grunted, turned on his heel and walked away. I followed him across the room. 'You may call it a filthy game' I said. 'Perhaps it was sometimes, but I knew then, as I know now, that we were cutting down the casualty lists and not piling them up. I may have done some very odd jobs but I am quite sure that what we did shortened the war and the slaughter'. This may have been what the General called a 'filthy game' but I am very glad to have been part of it and proud to tells its story."

To which I add ... me too!

Slava Ukraine!

Index

4th PsyOps Group (4th POG)   53, 230, 58
15 (UK) PsyOps Group   26, 46-47, 53, 84, 85
77th Brigade   48-49, 209, 231, 241

ABBA   214
Afghanistan   xi, xii, xiii, 6, 32, 47, 50, 53, 57, 58, 59, 63, 65, 66, 69, 70, 73-87, 90, 91, 124-126, 135, 137, 165, 168, 169, 184, 190, 195, 216, 220, 225, 233, 236, 246
Aggregate IQ   200
Al-Ahil Hospital   8
Al-Arabiya TV channel   5, 72
Al-Jazeera TV channel   5, 72-73, 74-75
Al-Qaeda   8, 32, 71, 72-73, 184
Al-Sadr, Muqtada   67
Afghan National Army (ANA)   62-64, 85
Anonymous   220-221, 226
Afghan National Police (ANP)   63-64, 81, 85
Afghan National Security Forces (ANSF)   62, 64, 82, 85
Advanced Research & Assessment Group (ARAG)   90-92, 126
Artificial Intelligence (AI)   viii, 155, 214-215, 222-224
Asia Foundation Polling   80-81

Baker Street (song)   181
Baseball   80
BBC World Service   169, 238
BBC Monitoring   238
BD methodology (BDM)   91, 92-93, 192, 210, 212-213, 217, 230, 235-236, 243, 245-246
Behavioural Conflict (book)   xi, 245
Behavioural Insights Team   244
Bell Pottinger   69-70, 164
Bellingcat   30, 115, 154, 186, 220, 222, 226
Biden, Joe President   17, 114, 136, 138, 180, 215
Big data   191-192, 197, 213
Billboards   68, 82, 97, 105
Black Lives Matter   143, 158
Blair, Tony Prime Minister   2, 43, 73, 107
BREXIT   xii, 7, 8, 22, 94, 153, 169, 170, 176-178, 183, 193, 194, 196, 199-200, 202, 204, 205, 206, 230
Briant, Emma Dr   196, 207, 208-209, 211, 212, 230
Buchardt, Jonny   161-162

Cadwalladr, Carole   192-194, 207
Cambridge Analytica   23, 94, 95, 190-213, 246
Carlson, Tucker   13, 134-135
Carroll, David Professor   195
Cats in the Cradle (Video)   46
Cyber & Electromagnetic Activities (CEMA)   26
CENTCOM   51, 73

250 Information Operations

Central Intelligence Agency
 (CIA)   55, 61, 122, 186, 194
Centreville incident   118
Chai Dawat (radio
 programme)   84
Chilcott Inquiry   15, 107
China   vii, viii, 8, 102, 103, 121,
 134, 135, 140, 141, 154-160, 183,
 217, 227, 232, 238, 239
Civil Military Cooperation
 (CIMIC)   65
Commander Joint PsyOps Task
 Force (CJPOTF)   79
Clinton, Hilary   16, 17, 87, 129,
 130, 205
Columbia   104-105
Commando Solo   108
Computational Propaganda   204,
 225
Conspiracy Theories   6, 15, 17, 28,
 62, 71, 174, 175-189
Cool Britannia   239
COVID   14, 48, 145, 157, 158, 182,
 187, 243
Cricket   80
Crimea   7, 121, 122, 126, 127, 128,
 129, 132
Cruz, Ted Senator   197
Cunningham, Finian   186-187
Cyber   vii, 26, 28, 50, 60, 96, 106,
 121, 122, 126, 127, 130, 141, 143,
 147, 150, 151-153, 155, 223, 227,
 229, 232, 236, 240, 247
Cyberberkut   105-106

Daly, Claire MEP   185
Dawn of Glad Tidings   108
De Quincey Adams, Jasper
 Brigadier   18-20
Deep State   14, 15, 16, 22, 176
Dehaye, Paul-Oliver Dr   194
Delmer, Sefton   35-37

Department for International
 Development (DFID)   101-103
Directorate Forward plans
 (DFP)   39-40
Directorate of Targeting and
 Information Operations
 (DTIO)   43
Disinformation   vii, viii, xiii, 20,
 22, 27-31, 118, 120, 147, 151, 159,
 192-193, 208, 213
Doctrine   2, 4, 23-24, 26, 28, 29,
 52, 53, 59, 92, 121, 125, 132, 146,
 203, 230
Defence Science Technology
 Limited (DSTL)   90, 93, 236
Dugin, Alexander   202

Easley, Matt General   60
Eastern Front Eunuchs   34-35
Ebola   119
Elections (Afghanistan)   59, 83,
 84, 87
Elections   105, 129, 140, 143, 153,
 157, 184, 191, 201, 218, 223
EMIC logic   91
Estonia   126, 127
Ethnocentrism   92
ETIC logic   92
Exercise CERBERUS 2020   18-19,
 22
Exercise JADE HELM   20

Facebook   193, 197, 200, 201, 204,
 214, 217, 218, 129, 132, 142, 144,
 152, 154, 156, 157, 169, 192, 194,
 198, 201-202, 205, 218-219
FARC   104-105
Foreign & Commonwealth Office
 (FCO)   101
Fifth Column (software)   223
Floyd, George   158, 165
Footballs   79, 167

Foreign Agents Registration
  Act   188
Foreign Influence Registration
  Scheme   187, 238
Fort Detrick   118
Freedom of Information
  (FOI)   208, 229, 10, 95, 193
Friedman, Laurence
  Professor   234

Galloway, George MP   185
Gaza   6, 8, 112-116
Government Communications
  Head Quarters (GCHQ)   243
Geospatial Intelligence
  (GEOINT)   66
Georgia   126, 209, 233
Giles, Kier   123, 237
Global Engagement Centre
  (GEC)   60-61
Goebbels, Joseph Dr   25, 216
Google   61, 80, 108, 151, 154, 158

Hamas   110-116
Hastings, Michael   66
Hayward, Tim Professor   186
Helmand Perception Matrix   81
Higgins, Eliot   185, 221
HIV / AIDS   90-91, 118-120
HMS Illustrious   1-3
Human Terrain Teams
  (HTS)   70-71, 92
Human Intelligence
  (HUMINT)   66
Hwang, Tim   196, 198

ICOS   76-78
Israeli Defence Forces
  (IDF)   110-113
Improvised Explosive Device
  (IED)   68-69, 84
Imagery Intelligence (IMINT)   66

Information Activities
  definition   27
Information Clearing
  House   186-187
Information Commissioner   191, 201
Information Professionals
  Association (IPA)   242
Information Research Department
  (IRD)   40-42
Information Warfare
  (definition)   26
Internet Research Agency
  (IRA)   117
Irish Republican Army (IRA)   140
Iran   8, 23, 67, 139-143, 145-147, 153, 160, 186, 208, 217, 231, 235, 240
Iraq   5, 6, 15, 29, 32, 50, 67-73, 107-110
Islamic Revolutionary Guard
  Corps (IRGC)   140, 141
International Security Assistance
  Force (ISAF)   79, 86
Islamic State (ISIS)   8, 108
Israel   110-111

Jamaica   96-101
Joint Improvised Explosive
  Device Defeat Organisation
  (JIEDDO)   68-69
Johnson, Boris   14-15
Jones, Alex   17, 20, 23, 180
Joint Terrorism Analysis Centre
  (JTAC)   75
Juba the Sniper   71-72

KISS principle   24
Kogan, Alexander Dr   191

Laity, Mark   194
Law of Armed Conflict   28, 29, 229

Lee's Life for Lies   72
Locus of Control   91, 173, 179, 248
London Controlling Station (LCS)   38-39
Long Range Gonad Reducer (LRGR)   85
Losing Arab Hearts and Minds (Book)   67
Loud Speakers   84
Luntz, Frank   87

Mackay, Andrew General   xi, xiii, 245
Mahdi Army   67
Mali   212, 216
Marine A   220
Marine Corps Information Operations Centre (MCIOC)   60
Maritime Intelligence (MARINT)   66
Martin, Mike Captain   xiii, 233
Maskirovka   124, 128
Maxwell, David Colonel   148, 153
McChrystal, Stanley General   65
McFate, Montgomery Professor   70, 71, 233
McKeigue, Paul Professor   186
McNeil Technologies   81
Measures of Effect (MOE)   49, 247
Media Ops (definition)   29
MH17   19, 30
MI5   243
Military Deception (MILDEC)   27, 29
Military Strategic Effects (MSE)   44, 103, 230
Miller, David Professor   186, 208-209
Miracle on the Hudson (Film)   ix, 232

Misinformation   27-28, 119, 194, 214
Military Information Support Operations (MISO)   26, 53
Military Strategic Effects (MSE)   103
Mugabe, Robert President   235
Mushy Peas   88-89
Musk, Elon   227
Myers, David Dr   165

Naler, Chris Colonel Dr   xiv
NATO   23, 63, 64
NATO Centre of Excellence Latvia   196, 210, 212-213, 245
NATO School Oberammergau   245
Neo-Taliban   75
Netherlands (Holland)   19, 64, 205, 212-213, 216
Night letters   85
Nix, Alexander   190-192, 195, 196, 197-198, 200, 206, 211
Normative Affiliation   217, 247
North Korea   ix, 8, 121, 147-153, 216
Norwegian Defence Research Agency (FFI)   194
NTrepid   218

Oakes, Nigel   xii, 191, 192, 196, 207, 246-247
Obama, Barack President   20, 62, 95, 117, 198, 215
OCEAN methodology   191, 197
Online Engagement Services   218
Op BASILICA   3, 4
Op CHRISTMAS   104
Op ENDURING FREEDOM   47, 63
Op HERRICK   63
Op IRAQI FREEDOM   63

Op KINGFISH 97-100
Op PALISER 4-5
Op PATWIN 103
Op SILKMAN 4-5
Op TELIC 63
Open Source Centre (OSC) 61
Operational Security (OPSEC) 27, 29, 30, 225
Open Source Intelligence (OSINT) 66, 224

Palantir 228
Pascale, Brad 11-12
People's Liberation Army (PLA) 154, 155
Philippines 101-103
Pizzagate 17
Political Warfare Executive (PWE) 34-38, 43, 55
Population Intelligence (POPINT) 66, 91, 95
Prigozhin, Yevgeny 117, 129
Propaganda definition 25
Provincial Reconstruction Teams (PRT) 65
PsyOps (PSYOP) 24, 26, 44-45, 52-54, 57, 61-62, 108
Public Affairs (definition) 29
Pulse (software) 219
Putin, Vladimir President vii, xii, 7, 19, 117, 121, 123, 125, 127, 129, 133-136, 137, 138, 183, 185, 186, 188, 202, 221, 223, 234, 235

QAnon 16, 18, 182

Radio 83-84, 108
Rand Corporation 59, 164, 165
Relationship Building Items (RBI) 79, 80, 97, 167
Reflexive Control 126, 128

Research & Information Communication Unit (RICU) 42, 210
Richards, Lee 36
Regulatory Investigative Powers Act (RIPA) 230
Rowland, Lee Dr 245
Royal Naval Reserve xiv, 240
RT TV 18
Revolutionary United Front (RUF) 1-3, 29
Rules of Engagement 3, 28, 29
Royal United Services Institute (RUSI) 66
Russia v-viii, xiii, xiv, 7, 9, 16, 18, 19, 22, 23, 30, 34, 62, 105, 106, 112, 117-138, 146, 153, 156-157, 159, 183, 187, 188, 193, 195, 202-204, 206, 210, 215-217, , 221, 231, 232, 234, 235, 238, 239, 240, 241

Sandyhook Massacre 17
Sarts, Janis 210
Sasquatch music festival 160-161
Saudi Arabia 79, 91
Schopenhauer, Arthur 245
SCL Defence 90, 93, 94, 190, 191-194, 195, 207, 208-210, 211, 212
Sharot, Tali Dr 169, 179
Shell 247
Sierra Leone 1-3, 6, 29-30
Signals Intelligence (SIGINT) 66
Space (domain) 51, 60, 97, 226-227, 228
Special Operations Executive (SOE) 33, 34
Sputnik 146, 187
Stalin, Joseph Leader of USSR 25, 163
Strategic Communication 24, 27, 67, 68, 91, 103, 113, 137, 169, 191, 195

Strava   225
Stuxnet   147
Suicide graduation ceremony   74
SVB (Bank) collapse   247
Sweden   21, 22, 136, 247
Syria   96, 108, 122, 145, 186-187, 226, 234

TAA   89, 92-93, 94, 192, 200, 210
Tahir Square   235
Taliban   59, 62, 64, 73, 79, 82, 85
Taylor, Phil Professor   108
The Big Lie (book)   248
The Marriage Bureau   76-78
Three Warfares doctrine   154
TikTok   14, 15, 137, 156, 159, 215
Trump, Donald President   7, 11-13, 15, 17, 22, 87, 94, 129-130, 133, 134, 143, 145, 148, 169, 176-177, 180, 181, 182, 191, 194, 197-199, 205-206, 242
Twitter (now X)   11, 13, 16, 19, 21, 22, 30, 109, 113, 115, 117, 119, 129, 136-137, 142, 144, 145, 158, 170, 186, 187, 194, 204, 205, 208, 215, 218-219, 221, 226, 240
Typhoon Haiyan   101-103

Ukraine   6, 8, 19, 21-22, 30, 32, 50, 62, 96, 105-107, 112-113, 117, 119, 121, 122, 123, 126, 127, 130-138, 157-159, 183, 186, 202, 203, 210, 218, 221, 222, 225, 226, 227, 228, 234-235, 239, 245, 247, 248

United Nations Mission to Sierra Leone (UNAMSIL)   2-3
United Arab Emirates (UAE)   115, 136, 145, 239
University of Maryland   241
Useful Idiots   120, 153, 185-188, 239, 242

Vanden Brook, Tom   70, 232, 233
Vets   84
Vietnam   57-58, 150, 178, 234, 247
Vk   105, 132, 156, 215

Wheatland, Julian   200
Whispers video   11-12, 22
Why we are here (DVD)   82
Williamson, Chris   186

YouTube   61, 154, 157, 158, 161, 164, 215, 218, 226

Zelensky, Volodymyr President   131, 136, 209

Previous Publications by the Author

Previous books

Tatham S A, Mackay A D, *Behavioural Conflict: Why Understand People's Motivations Will Prove Decisive in Future Conflict*, Military Studies Press, (2010).

Tatham S A, *Losing Arab Hearts & Minds: The Coalition, Al-Jazeera & Muslim Public Opinion*, Hurst & Co (2006), Front Street Press (2006).

Chapters in edited volumes

Tomchewis L, *Conflict in Urban Operations*, CHACR (2022).

Bains & O'Shaughnessy, *Propaganda*, SAGE (2011).

The British Army, *The British Army 2012: Warfare in the Fifth Dimensions*, Newsdesk Media (2012).

The British Army, *The British Army 2011: Information Operations. A Balance of Capabilities for an Unpredictable World*, Newsdesk Media (2011).

NATO, *Losing the Information War in Iraq: The Dynamics between Terrorism, Public Opinion & the Media*, NATO CoE, Ankara (2007).

Papers

Tatham S A, Tunnicliffe I, *Operationalizing Social Media in future US Information Operations*, US Army War College Strategic Studies Institute (Awaiting publication)

Tatham S A, Giles K, *Training Humans for the Human Domain*. US Army War College Strategic Studies Institute (2015).

Tatham S A, *NATO Audience Characterisation. The Potential for an Opponent to Exploit Fracture Lines in the NATO Audience*. NATO CoE Latvia (2015).

Tatham S A, *Using Target Audience Analysis to Aid Strategic Level Decisionmaking*, US Army War College Strategic Studies Institute (2015).

Tatham S A, *The Solution to Russian Propaganda Is Not EU or NATO Propaganda But Advanced Social Science to Understand and Mitigate Its Effect in Targeted Populations*, Latvian National Defence Academy (2015).

Tatham S A, Mackay A D, *Instability, Profitability, and Behavioral Change in Complex Environments*, Conflict Studies Research Centre, (2014).

Tatham S A, *U.S. Governmental Information Operations and Strategic Communications: A Discredited Tool or User Failure? Implications for Future Conflict*, Strategic Studies Institute US Army War College (2013).

Tatham S A, Le Page R, *NATO Strategic Communication: More to Do?* Published by National Defence Academy of Latvia (2014).

Tatham S A, *The Effectiveness of US military IO in Afghanistan: Why RAND Missed the Point.* UK Defence Academy, (2010).

Tatham S A, *Strategic Communication. An Important New Discipline.* The British Army Oct (2009).

Tatham S A, *Strategic Communication & Influence Operations: Do We Really Get it?* Published by UK Defence Academy (2009).

Tatham S A, Mackay A D, *Behavioural Conflict. From General to Strategic Corporal: Complexity, Adaptation and Influence.* UK Defence Academy (2009).

Tatham S A, *Strategic Communication: A Primer,* UK Defence Academy (2008).

Tatham S A, El-Katiri M, *Qatar: A Little Local Difficulty?* UK Defence Academy (2008).

Tatham S A, *Losing the Information War in Iraq,* NATO CoE Ankara (2007).

Tatham S A, *Al-Jazeera – Get Used to It. It's not Going Away.* Proceedings (2005).

Tatham S A, *Al-Jazeera: Can it make it here?* British Journalism Review, March (2005).

Tatham S A, *Operation Telic and the Media,* The Officer Magazine (2004).

Tatham S A, *Al-Jazeera. The Power of the Media,* The Officer Magazine (2004).

Tatham S A, *So What Did the Navy Ever do for Us?* The Sandy Times (2003).